SAI SPEED MATH ACADEMY

ABACUS MIND MATH

Excel at Mind Math with Soroban, a Japanese Abacus.

LEVEL – 1

WORKBOOK 1 OF 2

PUBLISHED BY SAI SPEED MATH ACADEMY

USA

www.abacus-math.com

Published in the United States of America by SAI Speed Math Academy, 2014

The Library of Congress has cataloged this book under this catalog number:
Library of Congress Control Number: 2014907005

ISBN of this edition: 978-1-941589-01-4

Printed in the United States of America

Edited by: WordPlay
www.wordplay.com

Front Cover Image: © [Yael Weiss] / Dollar Photo Club

www.abacus-math.com

Our Heartfelt Thanks to:

Our

Higher Self,

Family,

Teachers,

And Friends

for the support, guidance and confidence they gave us to…

…become one of the rare people who don't know how to quit. (~Robin Sharma)

KIND REQUEST

We believe knowledge is sacred.

We believe that knowledge has to be shared.

We could have monopolized our knowledge by franchising our work and creating wealth for ourselves. However, we choose to publish books so we can reach more parents and teachers who are interested in empowering their children with mind math at a very affordable cost and with the convenience of teaching at home.

Please help us know that we made the right decision by publishing books.

- We request that you please buy our books first hand to motivate us and show us your support.
- Please do not buy used books.
- We kindly ask you to refrain from copying this book in any form.
- Help us by introducing our books to your family and friends.

We are very grateful and truly believe that we are all connected through these books. We are very grateful to all the parents who have called in or emailed us to show their appreciation and support.

Thank you for trusting us and supporting our work.

With Best Regards,

SAI Speed Math Academy

Dear Parents and Teachers,

Thank you very much for buying this workbook. We are honored that you choose to use this workbook to help your child learn math and mind math using the Japanese abacus called the "Soroban". This is our effort to bring a much needed practice workbook to soroban enthusiasts around the world.

This book is the product of over six years of intense practice, research, and analysis of soroban. It has been perfected through learning, applying, and teaching the techniques to many students who have progressed and completed all six levels of our course successfully.

We are extremely grateful to all who have been involved in this extensive process and with the development of this book.

We know that with *effort, commitment* and *tenacity,* everyone can learn to work on soroban and succeed in mind math.

We wish all of you an enriching experience in learning to work on abacus and enjoying mind math excellence!

We are still learning and enjoying every minute of it!

GOAL AFTER COMPLETION OF LEVEL 1 – WORKBOOKS 1 AND 2

On successful completion of the two workbooks students would be able to:

1. Add any two digit numbers that does not involve carry-over or regrouping problems.
2. Subtract any two digit numbers that does not involve borrowing or regrouping problems.

HELPFUL SKILLS

- Know to read and write numbers 0-99
- Know to identify place value of numbers

INSTRUCTION BOOK FOR PARENTS/TEACHERS

Workbooks do not contain any instruction on how to work with an abacus. All instructions are in the Instruction Book for parents which is **sold separately** under the title:

Abacus Mind Math Level – 1 Instruction Book – ISBN: 978-1-941589-00-7

PRACTICE WORKBOOKS FOR STUDENTS

This book is Abacus Mind Math Level - 1 Workbook 1 of 2. Continue working with Abacus Mind Math Level 1 – **Workbook 2 of 2** after finishing this workbook to complete LEVEL – 1 training.
Workbook 2 of 2 is **sold separately** and is available under the title:
Abacus Mind Math Level – 1 Workbook 2 of 2 – ISBN: 978-1-941589-02-1

WE WOULD LIKE TO HEAR FROM YOU!

Please visit our Facebook page at https://www.facebook.com/AbacusMindMath.
Contact us through http://www.abacus-math.com/contactus.php or email us at **info@abacus-math.com**.

We Will Award Your Child a Certificate Upon Course Completion:

Once your child completes the test given at the back of the workbook – 2, please upload pictures of your child with completed test and marks scored on our Facebook page at https://www.facebook.com/AbacusMindMath, and at our email address: http://www.abacus-math.com/contactus.php

Provide us your email and we will email you a personalized certificate for your child. Please include your child's name as you would like for it to appear on the certificate.

LEARNING INSTITUTIONS AND HOME SCHOOLS

If you are from any public, charter or private school, and want to provide the opportunity of learning mind math using soroban to your students, please contact us. This book is a good teaching/learning aid for small groups or for one on one class. Books for larger classrooms are set up as 'Class work books' and 'Homework books'. These books will make the teaching and learning process a smooth, successful and empowering experience for teachers and students. We can work with you to provide the best learning experience for your students.

If you are from a home school group, please contact us if you need any help.

Contents

HOW TO USE THIS WORKBOOK

This workbook is for 10 weeks. Work in order given.

Each and every child is unique in his/her ability to learn. Sometimes a lesson might have to be repeated to get better understanding. You can erase the answers and redo the same lesson. On the other hand you may choose to move through the lessons quicker if child is an easy learner.

Each week's work is grouped together. Finish all the pages under each week before moving on to the next week's work. Work is divided for 5 days and you may choose to combine any number of days if you want to finish in shorter time. But, be very careful when you choose to do this, because children get overwhelmed easily when introduced to too many new concepts in a short time without having enough time to understand and practice. Use your best judgment since you know your child's temperament and learning capabilities.

After finishing a DAY's work check the answer and redo the problems with the wrong answer.

Wish you all the best!

INTRODUCTION

Sit at a desk with a comfortable height.
Study the picture and learn the names of the different parts of the abacus.
Practice clearing your abacus a few times.

FINGERING

Correct fingering is very important so, practice moving earth beads and heaven beads using the correct fingers.

WEEKS 1 to 4 – The goal for the first four weeks is to become familiar with the abacus and understand how a number, when set on abacus, looks. Children need to learn and understand that there are four earth beads and only one heaven bead.

WEEKS 5 to 7 – Mind math is introduced. Take time to make sure that the concept is well understood. Repeat each problem a couple of times to help with building skills.

WEEK 8 – Small friend combination facts are introduced.

WEEK 9 and 10 – Small friend formulas for +1, and -1 are introduced. First complete the rows that are to be worked with abacus. After this, move on to the section where it says to work in mind and try working on them in mind.

**** Continue working with ABACUS MIND MATH LEVEL 1 – WORKBOOK 2 OF 2 after finishing this workbook to complete LEVEL – 1 training.***

KEEPING TRACK

TIME: Make note of the time it takes your student to finish each day's work.

GRADES: Correct their work and calculate grade. Let your student color the stars next to each day's work. This will keep them engaged and encouraged.

GOAL: As student progresses through the week they should be able to do their work in less time with more accuracy.

HOW TO CALCULATE GRADE?

$$\frac{\text{Number of correct problems}}{\text{Total number of problems}} \times 100 = \text{Percentage scored}$$

GRADES

GRADE	PERCENTAGE	STAR COLOR
A+	96-100 EXTRAORDINARY	GOLD
A	91-95 EXCELLENT	
B+	86-90 AWESOME	SILVER
B	81-85 GOOD	
C	76-80 CONGRATULATIONS	BROWN
C+	70-75 CONGRATULATIONS	

PARTS OF ABACUS

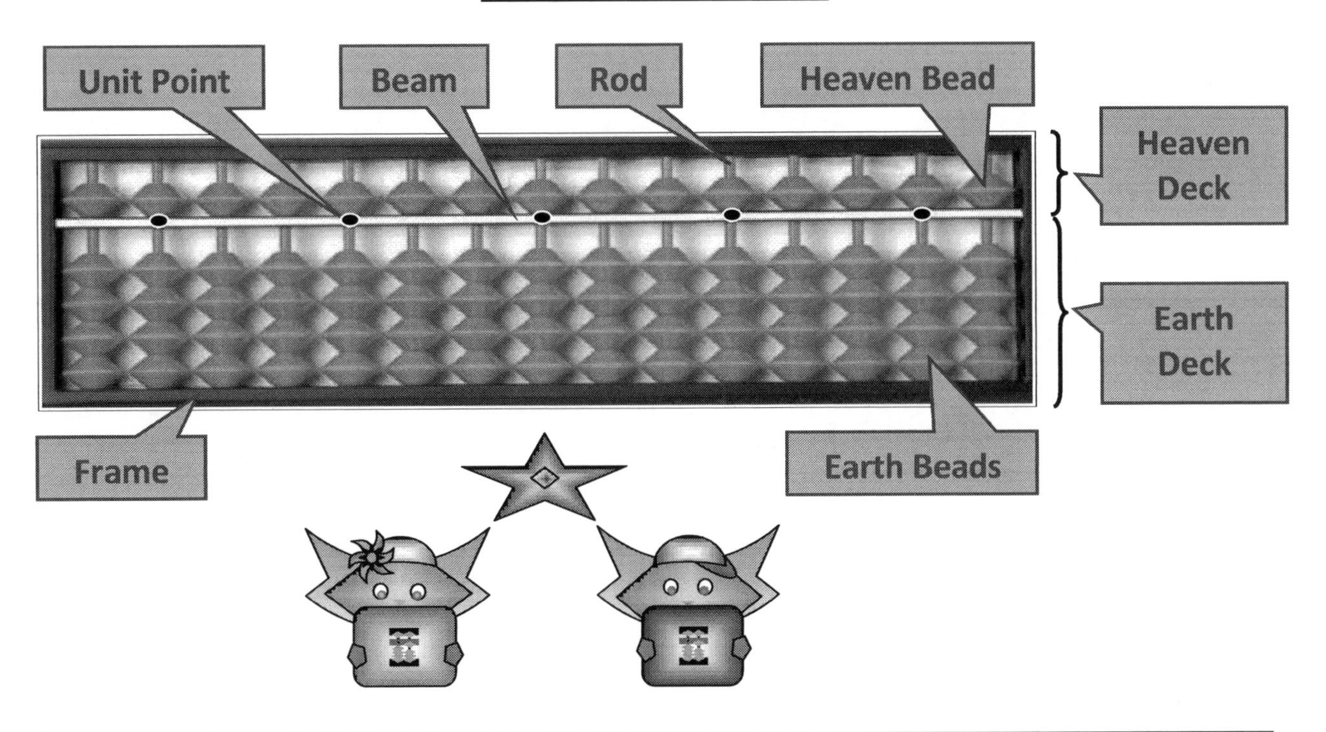

FRAME	Holds all the rods, beads and beam in place.
ROD	Sticks that hold the beads. Beads slide up or down on the rod.
BEADS	Represents numbers on the abacus. They slide on the rod and touch the beam or the frame. When beads touch the frame then your abacus is set at zero.
BEAM	The bar (usually white) that runs across all the rods and separates the Heaven and Earth beads. Only when beads touch the beam do they have value.
UNIT POINT	Can be used as decimal point. Can be used as comma that separates numbers by thousands. Example: $ 102,387,555 = One hundred and two million, three hundred eighty seven thousand, five hundred and fifty five dollars.
HEAVEN BEAD	There is one heaven bead above the beam on each rod. Each heaven bead is equal to "five".
EARTH BEADS	There are four earth beads below the beam on each rod. Each earth bead is equal to "one".

STUDENT'S SITTING POSITION

Sit at a table and place the abacus on the table with the heaven bead deck away from you. Please make sure that the table is not too high for the student.

CLEARING OR SETTING ABACUS AT ZERO

TRADITIONAL METHOD

STEP 1: Place the abacus on the table

STEP 2: Lift it with the bottom frame still touching the table. This will send all the earth beads to "zero" position

STEP 3: Gently place the abacus back on the table without moving the earth beads

STEP 4: Then place your finger between the heaven bead and the beam near the left hand side of the abacus

STEP 5: Drag your finger along the beam till to you reach the other side of the frame.

This will clear the heaven bead and send them to "zero" position

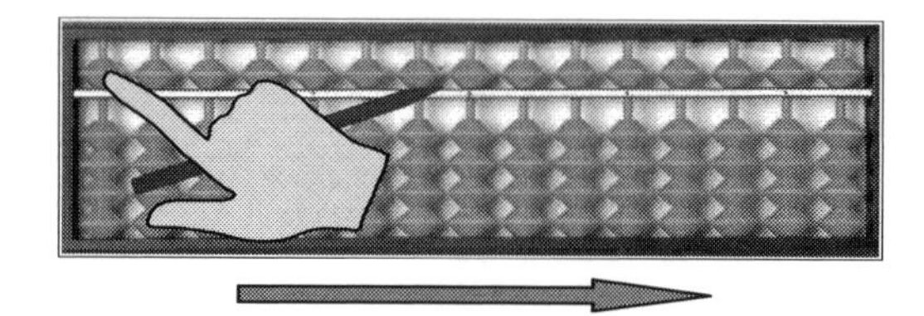

FUN METHOD: ZOOM AND CLEAR

STEP 1: Hold your thumb and pointer finger touching each other.

STEP 2: Place your fingers on the right side of the abacus beam with the beam in between the fingers like you are holding them very gently.

STEP 3: Now hold your abacus with the left hand so that the abacus does not move.

STEP 4: Now gently glide your fingers while still holding the beam, from the right side of the frame to the left side of the frame.

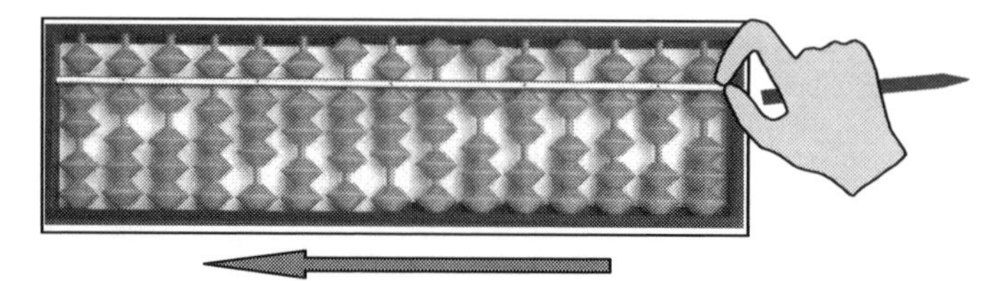

Clear your abacus every time you start a new calculation.

FINGERING

JOB OF THE THUMB (1):

A. Used to push the Earth Beads up to the beam, adding them to the game (ADD).

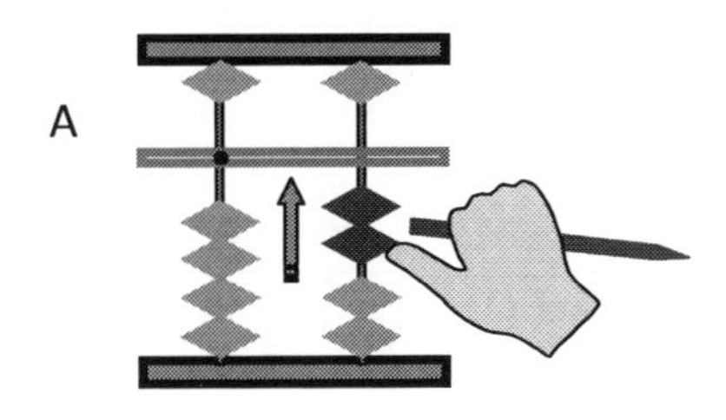

JOB OF THE POINTER FINGER (3):

1. Used to push the Earth Beads away from the beam, removing them from the game (MINUS).
2. Used to push the Heaven bead down to touch the beam, adding it to the game (ADD).
3. Used to push the Heaven bead away from the beam, removing it from the game (MINUS).

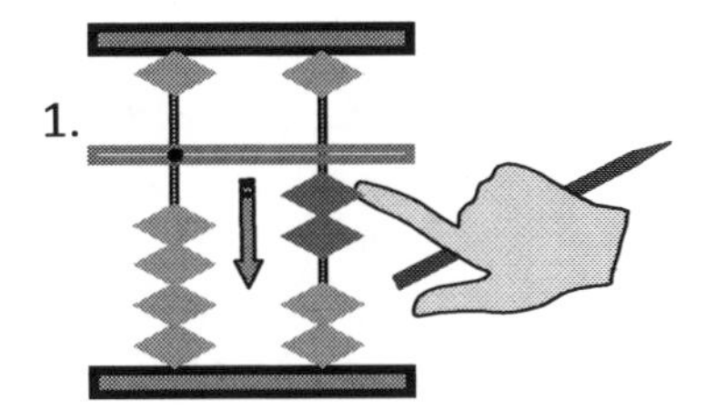

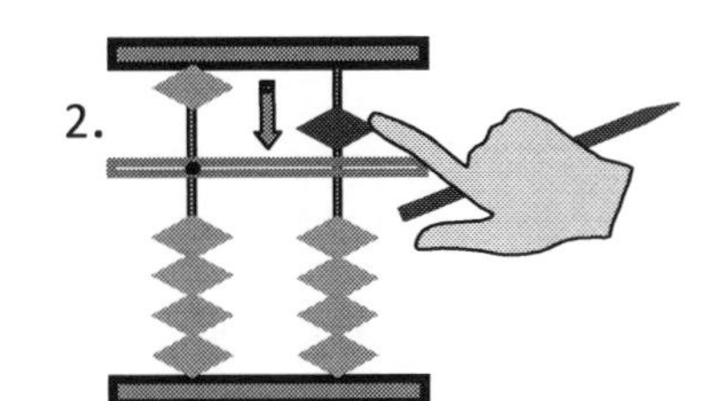

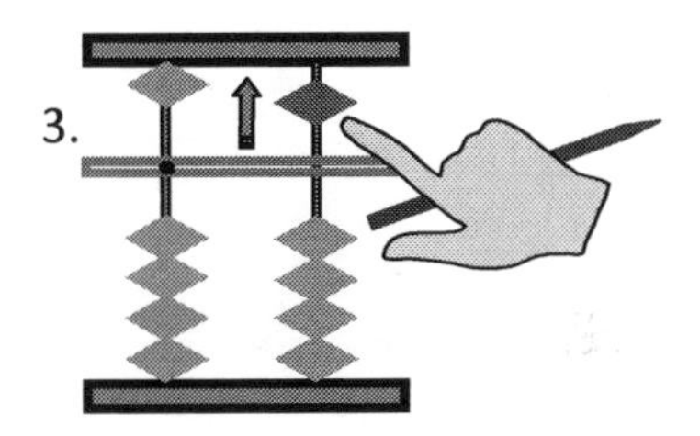

JOB OF THE OTHER THREE FINGERS:

Use your other three fingers to hold your pencil with the point facing down or away from you.

ADDING AND SUBTRACTING

ACTION FOR	ADDING	MINUS OR SUBTRACTING
EARTH BEADS	• When we say "adding" then it means we are moving the earth bead up to the beam with our thumb. • When the earth bead is touching the beam then it is "in the game" and is included in the reading. 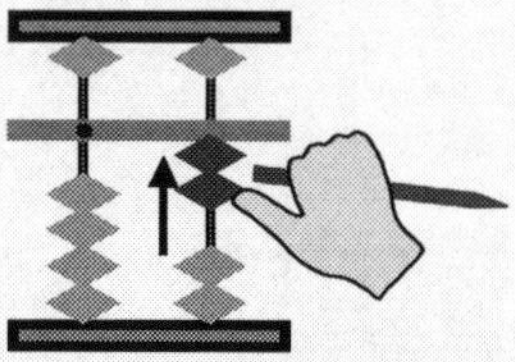	• When we say "minus" then it means we are moving the earth bead away from the beam and making it touch the frame or other beads that are not in the game with your pointer finger. • When the earth bead touches the frame then it means that the bead is "out of the game" and is not read. 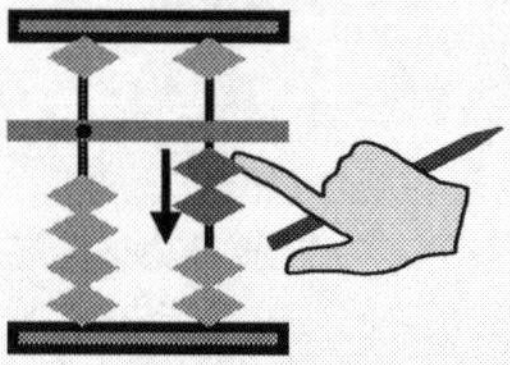
HEAVEN BEADS	• When we say "adding" it means we are moving the heaven bead with pointer finger to touch the beam. • When the heaven bead touches the beam it means that it is in the game and is included in the reading. 	• When we say "minus" then it means we are moving the heaven bead with our pointer finger away from the beam to make it touch the frame. • When the heaven bead touches the frame then it means that the bead is "out of the game" and is not read. 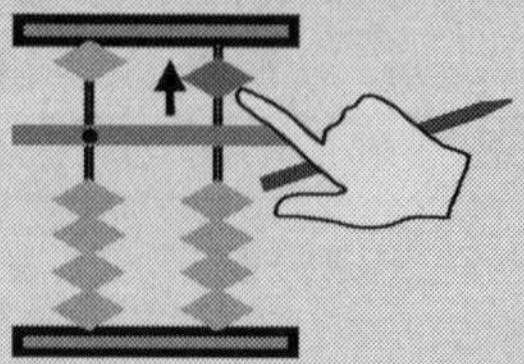

PLACE VALUES OF NUMBERS

Students need to know the names of the place value of numbers.

Example: 26

In the above number 2 is a tens place number and 6 is units place number.

PLACE VALUE PRACTICE

42 =	4	tens	2	ones	61 =	______	tens	______	ones
90 =	______	tens	______	ones	05 =	______	tens	______	ones
19 =	______	tens	______	ones	37 =	______	tens	______	ones
24 =	______	tens	______	ones	06 =	______	tens	______	ones
11 =	______	tens	______	ones	29 =	______	tens	______	ones
12 =	______	tens	______	ones	92 =	______	tens	______	ones
32 =	______	tens	______	ones	56 =	______	tens	______	ones
64 =	______	tens	______	ones	91 =	______	tens	______	ones
73 =	______	tens	______	ones	07 =	______	tens	______	ones
59 =	______	tens	______	ones	79 =	______	tens	______	ones
15 =	______	tens	______	ones	54 =	______	tens	______	ones
45 =	______	tens	______	ones	68 =	______	tens	______	ones
83 =	______	tens	______	ones	20 =	______	tens	______	ones

What place value is 2 in the following numbers? Write the name of the place value on the given line.

21 → Tens 23 → ________ 302 → ________

12 → ________ 62 → ________ 126 → ________

02 → ________ 24 → ________ 182 → ________

32 → ________ 52 → ________ 428 → ________

What place value is 5 in the following numbers? Write the name of the place value on the given line.

51 → ________ 56 → ________ 605 → ________

35 → ________ 85 → ________ 750 → ________

05 → ________ 57 → ________ 185 → ________

50 → ________ 95 → ________ 451 → ________

What place value is 7 in the following numbers? Write the name of the place value on the given line.

37 → ________ 47 → ________ 507 → ________

75 → ________ 07 → ________ 375 → ________

17 → ________ 79 → ________ 474 → ________

67 → ________ 27 → ________ 187 → ________

What place value is 9 in the following numbers? Write the name of the place value on the given line.

97 → ________ 49 → ________ 197 → ________

59 → ________ 19 → ________ 921 → ________

97 → ________ 89 → ________ 709 → ________

90 → ________ 92 → ________ 293 → ________

PLACE VALUE OF RODS

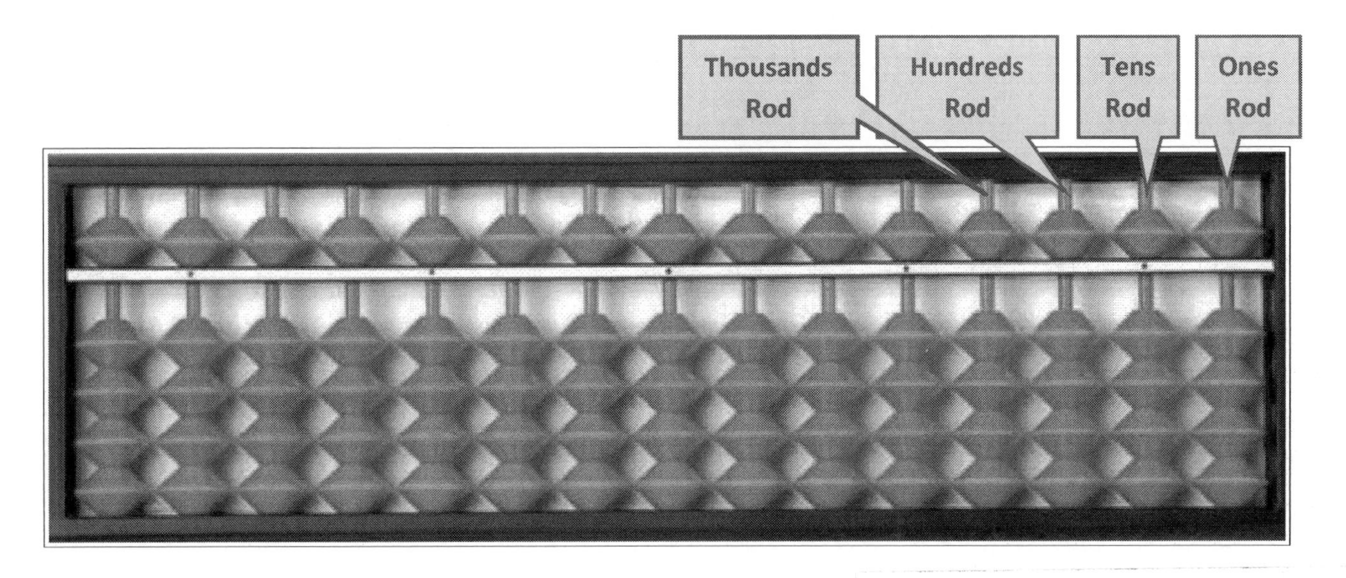

First rod from right is for the **ones place nu**
place number will be set on this

Second rod from right is for the **tens plac**
tens place number will be set on th

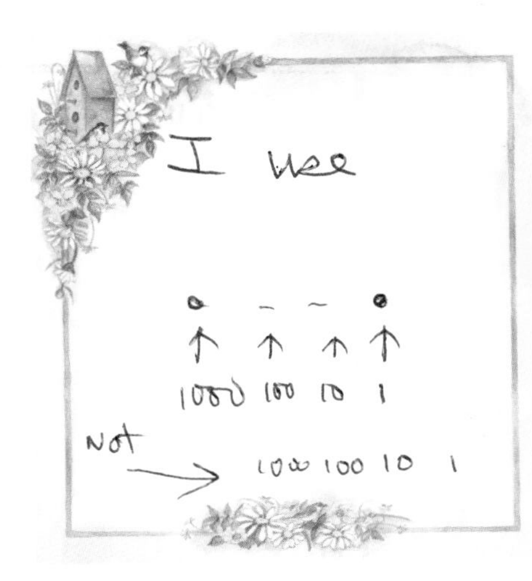

ORDER OF OPERATION

LEFT TO RIGHT: When working with two digit numbers: always add or subtract the tens place number first and then work on the ones place number.

WEEK 1 – LESSON 1 – INTRODUCING – EARTH BEADS

1) **Value of each earth bead = 1**
2) **There are four earth beads on each rod.**
3) **Use your thumb to move the earth bead up (adding) to touch the beam.**
4) **Use your pointer finger (index finger) to move the earth bead down (subtracting) to touch the frame.**
5) **Always set abacus to zero by clearing all the beads away from the beam before starting each calculation.**
6) **Setting numbers on the abacus:**

Tens place numbers go on the tens rod.

Ones place numbers go on the ones rod.

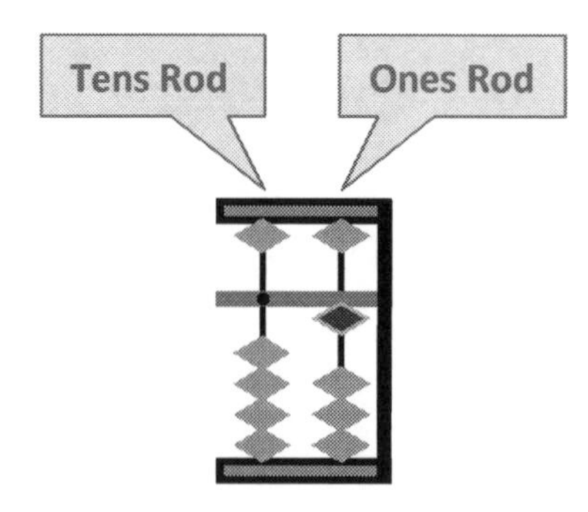

LESSON 1 – PRACTICE WORK

Use Abacus

DAY 1 – MONDAY

TIME: __________min __________sec Accuracy ______/40

1:1

1	2	3	4	5	6	7	8	9	10
01	01	01	02	02	03	03	04	11	13
01	02	03	02	01	01	- 02	- 01	02	- 02
01	01	- 01	- 01	01	- 02	03	- 01	01	11

1:2

1	2	3	4	5	6	7	8	9	10
14	43	22	12	33	24	10	30	20	30
- 12	- 32	21	32	- 21	- 10	20	10	20	- 10
21	01	01	- 41	- 11	- 10	10	- 20	- 10	20

1:3

1	2	3	4	5	6	7	8	9	10
				04	04	04	02	01	03
40	40	10	30	- 02	- 03	- 02	02	03	- 02
- 20	- 30	10	- 10	- 02	- 01	- 01	- 03	- 01	01
- 10	20	20	- 10	03	02	03	02	- 02	- 02

1:4

1	2	3	4	5	6	7	8	9	10
40	10	20	30	20	40	40	20	03	04
- 30	30	- 10	- 10	20	- 10	- 10	- 10	- 01	- 02
10	- 10	20	20	- 10	- 10	- 20	30	02	01
20	- 20	- 30	- 10	- 20	- 20	10	- 40	- 04	- 03

DAY 2 – TUESDAY

TIME: __________min __________sec Accuracy ______/16

1	2	3	4	5	6	7	8
31	33	31	42	21	11	33	43
12	01	- 10	01	13	21	11	- 12
- 41	- 04	- 10	- 33	- 14	- 32	- 23	01

1:5

1	2	3	4	5	6	7	8
14	13	22	31	42	40	24	24
- 03	30	- 10	- 10	02	- 10	- 01	- 10
11	- 12	22	22	- 10	04	- 03	30
02	- 20	- 33	- 11	- 20	- 23	11	- 44

1:6

DAY 3 – WEDNESDAY

TIME: __________min __________sec Accuracy ______/16

1	2	3	4	5	6	7	8
31	11	21	43	42	12	24	34
12	31	02	- 22	02	32	- 13	- 11
- 42	02	- 12	11	- 30	- 14	22	- 12
13	- 44	32	- 32	- 01	03	10	32

1:7

1	2	3	4	5	6	7	8
13	22	43	22	31	21	41	32
- 03	11	- 12	- 21	- 11	01	- 10	- 12
10	11	- 10	03	02	11	01	23
14	- 43	- 11	- 04	- 10	01	- 31	- 31

1:8

DAY 4 – THURSDAY

TIME: __________min __________sec Accuracy ______/16 ☆

1	2	3	4	5	6	7	8
23	44	43	31	21	22	12	41
- 03	- 31	- 21	13	12	11	12	- 30
22	21	22	- 22	11	- 13	- 02	03
- 02	- 31	- 01	- 11	- 34	24	- 10	- 11

1:9

1	2	3	4	5	6	7	8
11	02	12	11	31	31	32	32
11	10	02	12	01	11	- 11	- 11
11	10	10	01	- 21	- 41	21	- 11
11	22	10	20	20	11	- 10	31

1:10

DAY 5 – FRIDAY

TIME: __________min __________sec Accuracy ______/16 ☆

1	2	3	4	5	6	7	8
11	22	22	11	11	12	22	32
01	01	02	12	11	11	- 11	- 10
01	01	10	11	- 12	- 13	22	- 10
11	10	10	10	22	12	- 11	20

1:11

1	2	3	4	5	6	7	8
31	41	11	13	22	32	44	21
02	01	12	- 02	12	02	- 03	22
- 22	02	- 03	11	- 33	- 04	02	- 33
01	- 21	12	- 21	11	13	- 40	14

1:12

LESSON 1 – SKILL BUILDING EXERCISE

Set the given numbers on the abacus. See where the beads are on the abacus.
Copy that by drawing beads on the abacus picture below to show the numbers.

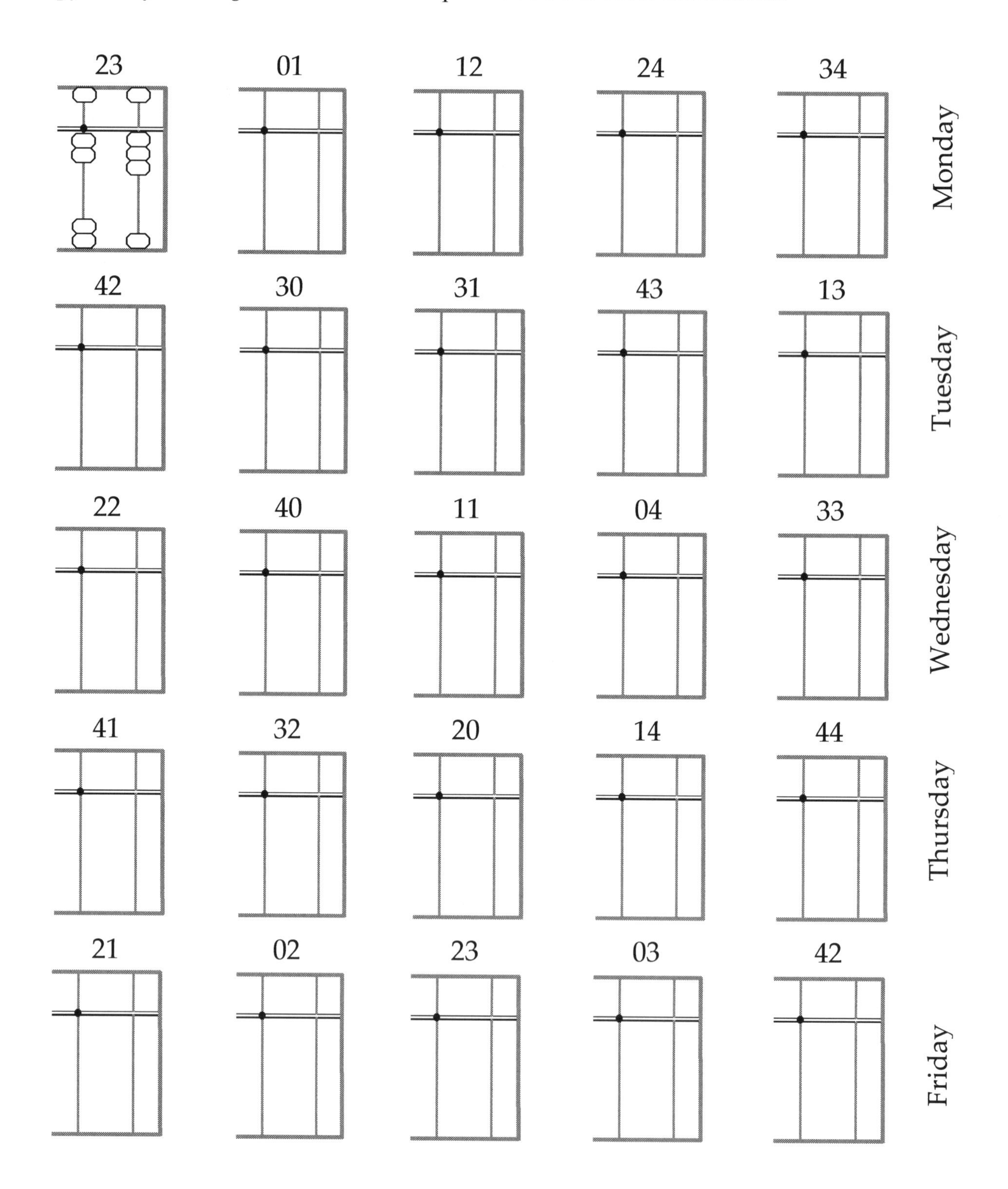

WEEK 2 – LESSON 2 – INTRODUCING – HEAVEN BEAD

1) Value of heaven bead = 5

2) There is one heaven bead above the beam (heaven deck) on each rod.

3) Use your pointer finger (index finger) to move the heaven bead down (adding) to touch the beam.

4) Use your pointer finger (index finger) to move the heaven bead up (subtracting) to touch the frame.

5) Always set abacus to zero by clearing all the beads away from the beam before starting each calculation.

6) Setting numbers on the abacus:

Tens place numbers go on the tens rod.

Ones place numbers go on the ones rod.

LESSON 2 – PRACTICE WORK

Use Abacus

DAY 1 – MONDAY

TIME: ________min ________sec Accuracy ______/40

1	2	3	4	5	6	7	8	9	10
25	02	02	11	05	11	32	22	31	11
10	51	02	02	30	15	- 21	- 11	52	11
10	- 03	50	11	- 10	- 01	52	- 11	- 50	12
- 05	05	- 04	10	- 05	10	- 50	45	11	- 21

2:1

1	2	3	4	5	6	7	8	9	10
51	23	12	52	24	33	42	43	55	53
22	- 22	35	32	- 13	- 22	05	- 11	- 05	- 01
- 12	45	- 01	- 54	35	05	01	- 21	04	15
- 51	- 35	- 01	05	- 05	- 11	- 35	- 11	- 50	- 12

2:2

1	2	3	4	5	6	7	8	9	10
51	22	42	53	54	54	45	11	25	02
02	- 11	- 22	- 02	- 03	- 01	53	12	51	02
11	25	53	05	05	- 51	- 31	51	11	50
- 14	- 11	- 21	- 51	- 55	22	- 15	- 23	- 32	- 01

2:3

1	2	3	4	5	6	7	8	9	10
11	55	32	42	52	55	42	53	51	42
12	02	02	51	11	12	51	- 52	15	51
11	- 51	- 11	01	21	01	01	31	11	01
10	- 01	- 11	- 44	- 54	- 53	- 52	- 11	- 52	- 42

2:4

DAY 2 – TUESDAY

TIME: ___________min ___________sec Accuracy _______/16 ☆

1	2	3	4	5	6	7	8
35	21	22	14	54	52	11	52
12	25	22	55	- 01	- 01	32	41
- 01	02	- 14	- 01	- 01	31	05	- 31
- 15	- 33	05	- 53	- 01	- 51	- 45	- 12

2:5

1	2	3	4	5	6	7	8
15	42	22	51	32	51	53	52
32	05	22	03	15	32	- 02	02
- 30	- 21	- 33	- 52	- 22	- 53	- 51	- 03
- 05	- 21	11	51	- 05	05	05	- 51

2:6

DAY 3 – WEDNESDAY

TIME: ___________min ___________sec Accuracy _______/16 ☆

1	2	3	4	5	6	7	8
41	12	12	11	35	21	15	53
01	15	52	32	11	12	11	- 01
05	01	15	05	- 41	- 31	- 25	- 02
- 02	- 15	- 54	- 33	50	51	11	05

2:7

1	2	3	4	5	6	7	8
15	25	51	23	53	52	24	35
13	01	22	- 01	- 02	- 01	- 03	10
- 25	22	- 53	25	01	45	11	- 25
51	- 43	05	- 12	- 50	- 51	- 31	15

2:8

DAY 4 – THURSDAY

TIME: __________min ____________sec Accuracy ______/16 ☆

1	2	3	4	5	6	7	8	2:9
55	53	25	12	24	53	32	54	
32	- 02	23	11	- 03	- 02	12	- 01	
- 52	35	- 45	- 01	25	15	55	- 01	
- 05	- 51	51	10	- 01	- 11	- 45	- 52	

1	2	3	4	5	6	7	8	2:10
54	44	54	15	52	15	22	22	
- 51	- 22	- 02	14	15	23	25	21	
45	55	05	- 02	21	- 35	52	- 42	
- 33	- 22	- 55	- 05	- 55	51	- 44	50	

DAY 5 – FRIDAY

TIME: __________min ____________sec Accuracy ______/16 ☆

1	2	3	4	5	6	7	8	2:11
25	53	15	55	22	53	25	15	
- 10	- 01	30	- 05	11	11	52	11	
30	05	- 15	02	15	- 01	- 12	- 01	
- 15	- 52	10	- 02	- 03	- 52	- 15	- 20	

1	2	3	4	5	6	7	8	2:12
35	11	33	54	31	35	25	13	
12	01	10	- 01	13	11	13	50	
- 25	- 02	- 03	15	- 22	- 41	50	- 03	
01	35	05	- 53	- 20	50	- 35	- 50	

LESSON 2 – SKILL BUILDING EXERCISE

Set the given numbers on the abacus. See where the beads are on the abacus.
Copy that by drawing beads on the abacus picture below to show the numbers.

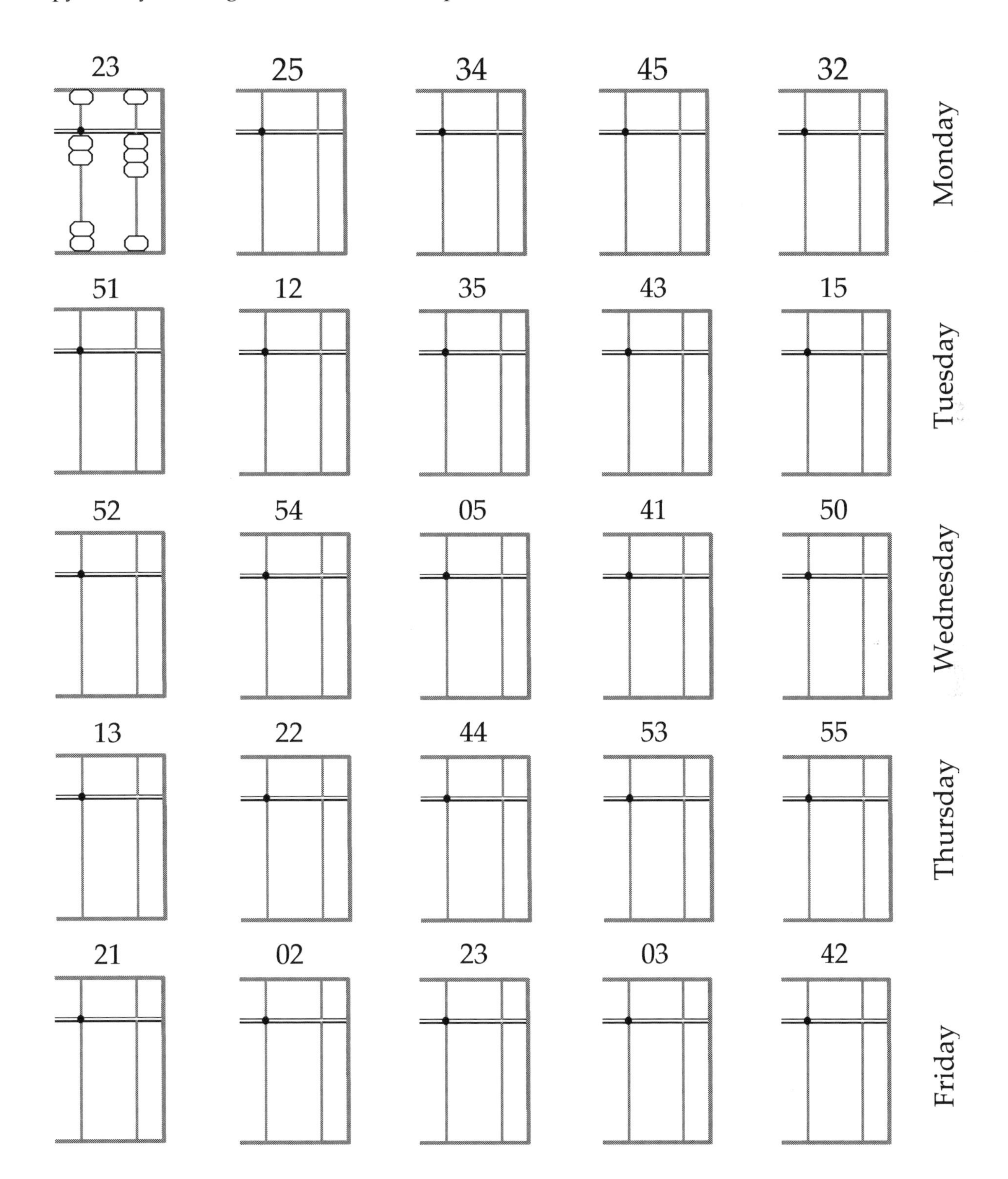

WEEK 3 – LESSON 3 – INTRODUCING– NUMBERS 6 AND 7

Use Abacus

LESSON 3 – PRACTICE WORK

DAY 1 – MONDAY

TIME: ________min ________sec Accuracy ______/40 ☆

1	2	3	4	5	6	7	8	9	10	
03	05	07	02	07	02	07	03	04	07	3:1
- 01	02	- 01	05	- 02	- 01	- 05	- 01	- 01	01	
- 01	- 02	- 01	- 01	01	05	- 01	05	05	- 03	
05	01	02	01	- 05	- 01	06	- 02	- 01	02	

1	2	3	4	5	6	7	8	9	10	
20	20	60	40	10	01	04	20	20	04	3:2
- 10	- 20	- 10	- 20	20	03	- 03	10	20	- 02	
50	10	20	50	- 10	- 02	05	06	06	05	
10	50	- 10	- 10	50	05	01	01	01	- 06	

1	2	3	4	5	6	7	8	9	10	
56	16	42	30	50	10	51	21	12	25	3:3
11	- 05	01	- 20	30	10	21	01	10	51	
- 01	12	01	50	- 10	- 20	- 10	05	10	- 10	
- 15	51	- 32	10	- 10	70	- 10	- 11	05	- 60	

1	2	3	4	5	6	7	8	9	10	
20	65	56	76	21	67	25	07	60	60	3:4
20	12	21	- 20	10	10	10	- 02	- 50	- 10	
- 30	- 56	- 10	11	16	- 20	11	- 05	60	20	
60	25	- 17	- 57	- 40	- 56	- 20	50	- 20	- 70	

DAY 2 – TUESDAY

TIME: __________min __________sec Accuracy ______/16

1	2	3	4	5	6	7	8
05	26	03	50	22	40	42	25
12	- 05	- 01	20	05	- 30	- 30	- 10
- 02	02	- 01	- 60	21	50	51	51
21	01	05	50	- 01	- 10	- 13	- 10

3:5

1	2	3	4	5	6	7	8
31	26	65	22	12	44	74	25
- 20	51	12	55	55	55	- 02	- 20
55	- 72	- 01	- 12	- 17	- 11	- 02	12
- 10	01	- 50	- 05	26	- 11	06	50

3:6

DAY 3 – WEDNESDAY

TIME: __________min __________sec Accuracy ______/16

1	2	3	4	5	6	7	8
21	12	25	20	07	10	57	15
21	50	21	10	01	02	30	10
05	- 10	- 06	06	- 03	50	- 70	- 20
- 11	25	01	01	02	01	- 11	71

3:7

1	2	3	4	5	6	7	8
50	21	72	15	74	61	62	25
30	56	- 11	51	- 21	13	12	51
- 10	- 10	15	- 16	- 03	- 24	- 01	- 10
- 20	- 60	- 55	27	06	12	- 21	- 16

3:8

DAY 4 – THURSDAY

TIME: ________ min ________ sec Accuracy ______ /16

1	2	3	4	5	6	7	8	
35	45	43	62	77	65	22	12	3:9
- 15	51	01	12	- 11	12	55	55	
02	- 50	- 21	- 73	21	- 01	- 67	- 11	
22	- 10	50	51	- 50	- 50	54	20	

1	2	3	4	5	6	7	8	
11	74	41	25	12	21	33	67	3:10
55	- 02	05	20	55	22	- 11	- 55	
- 11	02	- 35	- 35	- 61	- 13	- 11	15	
22	- 61	60	16	71	04	65	- 22	

DAY 5 – FRIDAY

TIME: ________ min ________ sec Accuracy ______ /16

1	2	3	4	5	6	7	8	
65	53	25	75	45	43	62	66	3:11
- 10	- 02	52	- 25	52	01	12	11	
21	15	- 15	11	- 50	- 21	- 73	- 20	
- 55	- 50	05	06	- 16	50	66	- 07	

1	2	3	4	5	6	7	8	
35	51	32	54	31	35	25	23	3:12
12	31	15	- 02	13	13	13	50	
- 26	- 02	- 30	15	- 22	- 42	- 25	- 11	
55	- 50	50	- 06	51	51	51	15	

LESSON 3 – SKILL BUILDING EXERCISE

Set the given numbers on the abacus. See where the beads are on the abacus.
Copy that by drawing beads on the abacus picture below to show the numbers.

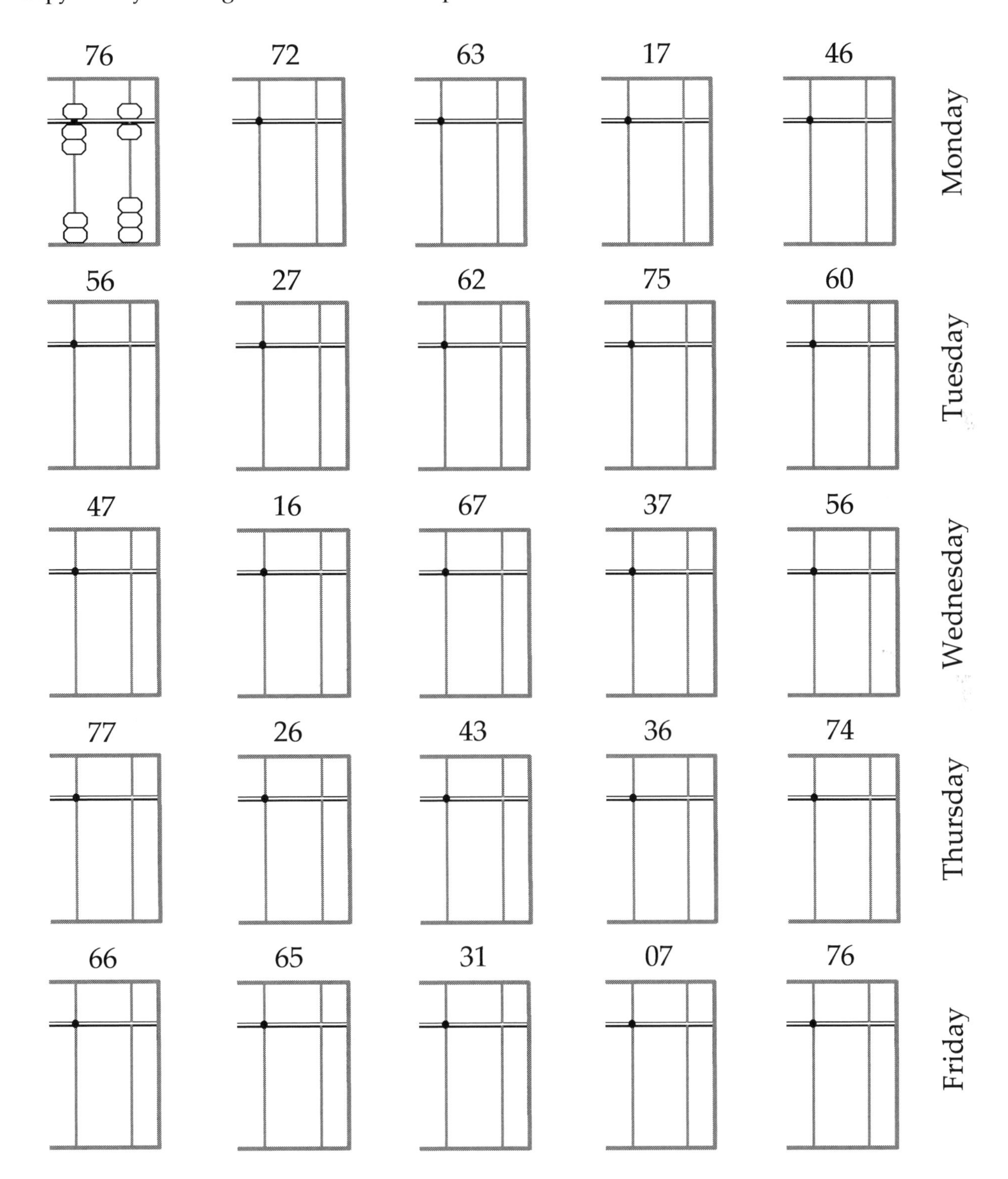

WEEK 4 – LESSON 4 – INTRODUCING – NUMBERS 8 AND 9

Use Abacus

LESSON 4 – PRACTICE WORK

DAY 1 – MONDAY

TIME: ____________min ______________sec Accuracy ________/40 ☆

1	2	3	4	5	6	7	8	9	10
04	40	01	07	10	20	70	22	45	20
- 03	- 10	01	01	20	- 10	- 10	05	- 05	20
02	60	06	- 02	10	50	30	21	03	06
05	- 10	01	03	50	- 10	- 10	01	05	01

4:1

1	2	3	4	5	6	7	8	9	10
08	08	09	09	41	30	50	30	50	40
01	- 03	- 01	- 03	01	- 20	30	10	30	- 20
- 02	02	- 01	02	02	50	- 10	- 20	- 10	50
01	01	02	01	- 31	10	20	70	- 20	10

4:2

1	2	3	4	5	6	7	8	9	10
08	09	04	52	34	13	60	90	70	20
- 02	- 04	05	12	- 03	25	20	- 20	10	10
03	02	- 01	- 04	05	- 30	- 10	- 10	10	10
- 01	02	- 02	05	51	60	20	- 10	- 50	- 40

4:3

1	2	3	4	5	6	7	8	9	10
25	42	43	62	75	61	21	12	44	74
- 15	55	01	12	11	12	75	12	55	- 01
02	- 11	- 22	- 03	- 21	- 01	02	- 04	- 11	- 01
12	03	75	02	33	- 51	- 83	15	- 11	- 01

4:4

DAY 2 – TUESDAY

TIME: __________min ____________sec Accuracy _______/16 

1	2	3	4	5	6	7	8	
07 - 02 - 05 08	02 01 05 - 03	09 - 01 - 01 - 01	80 - 30 20 20	60 10 10 - 70	01 01 05 01	72 15 11 - 75	82 02 15 - 67	4:5

1	2	3	4	5	6	7	8	
91 02 05 - 82	55 41 - 91 82	41 52 - 82 05	75 24 - 95 45	32 55 - 61 53	21 62 - 71 67	83 - 11 26 - 85	19 - 02 - 05 75	4:6

DAY 3 – WEDNESDAY

TIME: __________min ____________sec Accuracy _______/16

1	2	3	4	5	6	7	8	
03 - 01 05 01	09 - 02 01 - 08	20 50 20 - 10	03 - 01 06 01	90 - 70 20 50	57 42 - 85 10	88 - 06 - 72 05	43 55 - 92 01	4:7

1	2	3	4	5	6	7	8	
52 46 - 48 20	95 - 40 11 23	85 - 30 44 - 20	67 12 - 50 - 29	14 - 02 05 - 11	57 - 05 22 - 03	79 - 23 42 - 87	36 53 - 18 27	4:8

DAY 4 – THURSDAY

TIME: __________min ____________sec Accuracy _______/16

1	2	3	4	5	6	7	8
07	60	90	80	75	23	30	39
- 05	- 10	- 10	10	12	20	- 10	- 28
- 01	40	- 10	- 50	- 30	- 30	50	17
08	- 10	- 50	- 30	21	56	- 10	50

4:9

1	2	3	4	5	6	7	8
73	24	09	85	22	94	27	21
- 12	- 03	- 03	04	11	- 30	11	58
35	75	01	- 12	10	- 12	51	- 15
- 01	- 11	01	- 01	51	20	- 18	30

4:10

DAY 5 – FRIDAY

TIME: __________min ____________sec Accuracy _______/16

1	2	3	4	5	6	7	8
71	80	36	53	11	72	81	62
12	- 50	52	- 01	52	21	13	12
15	10	- 67	12	11	- 01	- 24	- 01
- 40	50	18	05	15	- 02	08	- 21

4:11

1	2	3	4	5	6	7	8
51	05	53	21	23	43	16	39
05	64	25	01	15	01	22	- 01
01	- 01	- 10	21	51	55	01	- 25
22	- 51	- 10	55	- 02	- 32	- 03	- 01

4:12

LESSON 4 – SKILL BUILDING EXERCISE

Set the given numbers on the abacus. See where the beads are on the abacus.
Copy that by drawing beads on the abacus picture below to show the numbers.

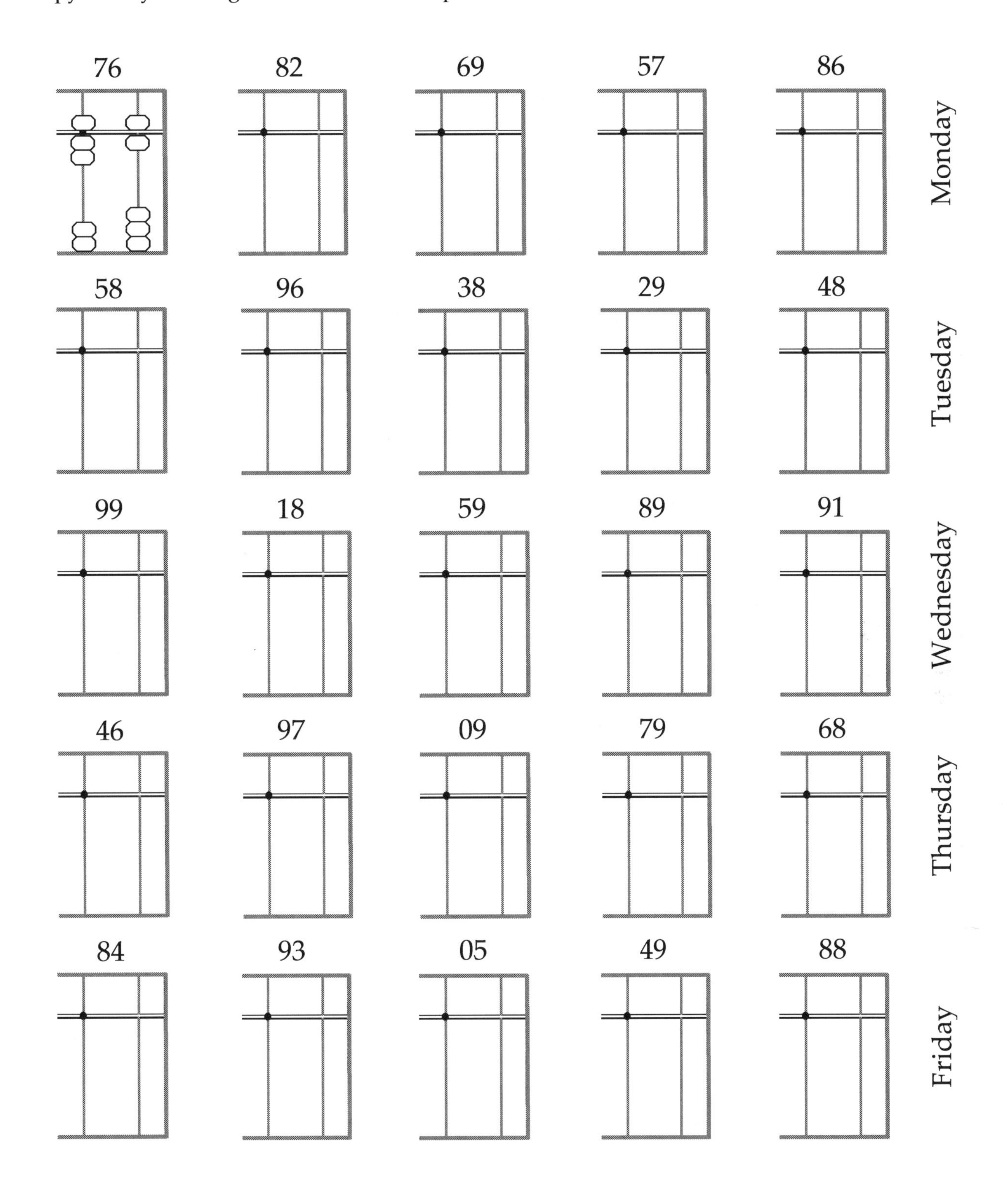

WEEK 5 – LESSON 5 – INTRODUCING – MIND MATH

LESSON 5 – PRACTICE WORK

Use Abacus

DAY 1 – MONDAY

TIME: ________ min ________ sec Accuracy ______/16

1	2	3	4	5	6	7	8
11	51	25	13	24	11	10	24
10	- 01	63	- 03	- 03	21	27	25
23	35	01	85	72	55	- 16	50
05	- 05	- 82	- 05	- 33	10	27	- 10

5:1

1	2	3	4	5	6	7	8
10	33	12	22	61	21	28	65
20	11	30	77	- 60	68	71	24
59	50	55	- 66	41	- 57	- 89	- 59
10	- 40	- 86	- 31	- 40	66	76	18

5:2

DAY 2 – TUESDAY

TIME: ________ min ________ sec Accuracy ______/16

1	2	3	4	5	6	7	8
53	65	55	24	95	38	20	35
10	04	03	20	04	- 17	25	52
01	- 55	01	50	- 02	26	50	10
30	20	- 54	- 02	- 71	50	- 35	- 50

5:3

1	2	3	4	5	6	7	8
29	41	38	50	21	39	54	63
20	52	11	13	21	- 27	05	35
- 39	- 11	- 12	15	- 40	75	- 56	- 96
78	- 11	- 10	20	51	02	35	72

5:4

DAY 3 – WEDNESDAY

TIME: ________min ________sec Accuracy ______/16 ☆

1	2	3	4	5	6	7	8
22	64	26	84	56	53	25	25
55	- 01	63	- 53	- 01	15	24	- 15
22	25	- 58	18	33	20	- 05	53
- 68	- 75	- 31	- 25	11	- 12	50	- 11

5:5

1	2	3	4	5	6	7	8
84	10	17	14	20	12	32	27
10	04	21	25	24	56	12	52
05	10	- 18	- 12	55	11	55	- 15
- 72	05	10	21	- 18	- 17	- 49	- 64

5:6

DAY 4 – THURSDAY

TIME: ________min ________sec Accuracy ______/16 ☆

1	2	3	4	5	6	7	8
56	43	20	79	18	10	20	47
11	- 21	63	- 63	- 07	35	02	- 31
21	65	01	51	11	54	77	52
- 85	- 75	05	- 65	25	- 55	- 84	- 15

5:7

1	2	3	4	5	6	7	8
21	37	10	21	38	21	35	33
12	10	30	22	- 20	11	10	11
51	50	55	55	50	55	52	55
15	- 26	- 50	- 12	- 16	10	- 25	- 98

5:8

DAY 5 – FRIDAY

TIME: __________min __________sec Accuracy ______/16

1	2	3	4	5	6	7	8
24	20	10	19	21	32	27	41
25	24	56	- 16	25	52	21	52
- 17	50	10	21	52	10	- 37	- 11
51	- 60	11	55	- 87	- 50	50	- 11

5:9

1	2	3	4	5	6	7	8
37	36	50	25	39	50	39	43
10	11	13	23	- 27	01	- 36	- 12
50	- 12	15	- 47	25	- 50	55	51
- 26	- 10	- 56	26	62	12	- 56	16

5:10

LESSON 5 – SKILL BUILDING EXERCISE

Draw beads to show numbers given.

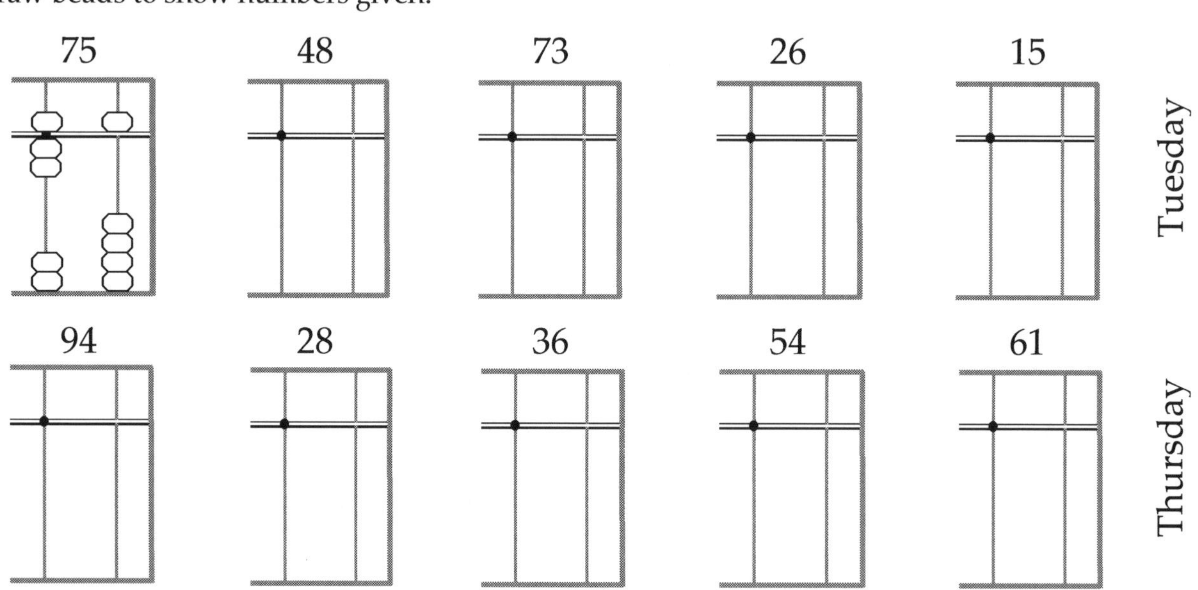

LESSON 5 – MIND MATH PRACTICE

Visualize 

DAY 1 – MONDAY

Accuracy ______/40 ☆

1	2	3	4	5	6	7	8	9	10	5:11
01	02	03	04	03	04	02	03	01	04	
02	01	01	- 01	- 02	- 02	02	- 01	03	- 03	

1	2	3	4	5	6	7	8	9	10	5:12
					01	20	30	12	23	
10	20	30	10	20	02	10	10	- 10	- 02	
10	20	- 10	20	- 10	01	10	- 20	20	- 01	

1	2	3	4	5	6	7	8	9	10	5:13
41	31	14	22	33	33	21	34	34	44	
02	10	- 02	02	01	- 20	10	- 20	- 02	- 33	
- 01	- 10	- 02	- 01	- 02	- 10	01	- 01	- 10	10	

1	2	3	4	5	6	7	8	9	10	5:14
05	40	15	30	15	54	51	03	10	04	
30	05	20	05	10	- 03	03	50	10	50	
- 10	- 20	10	- 20	- 20	- 51	- 04	- 01	05	- 03	

DAY 2 – TUESDAY

Accuracy ______/10 ☆

1	2	3	4	5	6	7	8	9	10	5:15
11	22	22	12	44	11	10	05	20	10	
11	11	22	11	- 22	01	20	10	05	05	
11	11	- 11	10	- 12	20	05	10	- 10	30	

DAY 3 – WEDNESDAY

Accuracy ______/10 ☆

1	2	3	4	5	6	7	8	9	10	5:16
41	31	23	15	10	30	52	50	05	20	
01	12	21	20	05	05	- 01	05	10	05	
02	- 41	- 32	- 05	- 10	- 30	- 01	- 50	- 05	- 25	

DAY 4 – THURSDAY

Accuracy ______/10 ☆

1	2	3	4	5	6	7	8	9	10	5:17
33	33	40	20	40	50	45	05	05	54	
- 11	11	- 30	20	- 10	02	- 30	30	20	- 03	
- 22	- 22	05	- 30	- 20	- 50	- 10	10	- 15	02	

DAY 5 – FRIDAY

Accuracy ______/20 ☆

1	2	3	4	5	6	7	8	9	10	5:18
55	02	54	45	35	55	20	55	10	44	
- 05	52	- 01	- 40	- 05	- 50	25	- 50	25	- 33	
01	- 03	- 02	50	15	30	- 45	20	- 05	- 11	

1	2	3	4	5	6	7	8	9	10	5:19
50	20	20	55	30	20	44	33	44	44	
02	10	15	- 50	- 10	15	- 22	- 22	- 14	- 34	
02	05	- 30	10	25	- 25	11	11	05	15	

LESSON 5 – SKILL BUILDING

Visualize the numbers on the beam in your mind and draw it to represent the numbers given.

12	35	05	15	25	
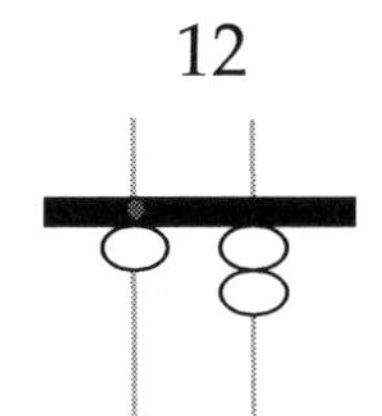		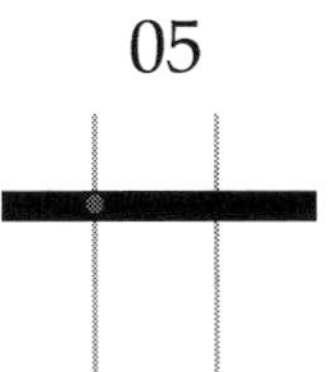		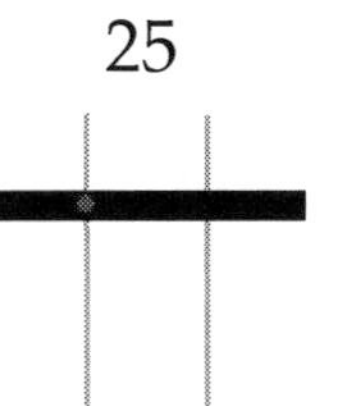	Monday
33	**43**	**51**	**11**	**54**	
	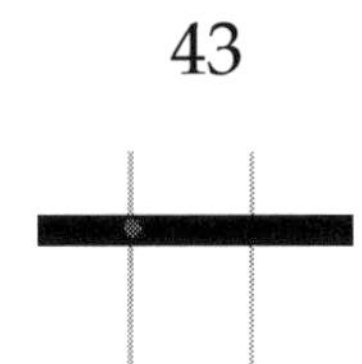	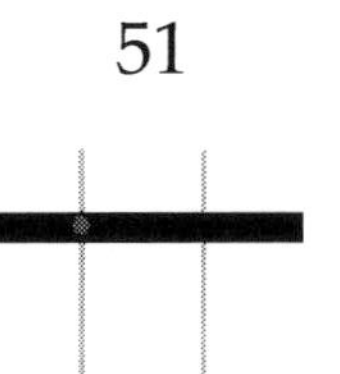	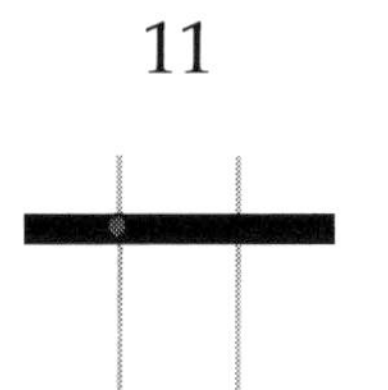		Tuesday
32	**21**	**31**	**45**	**24**	
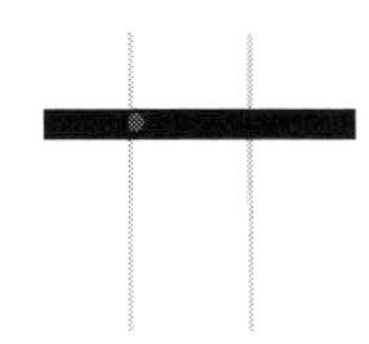	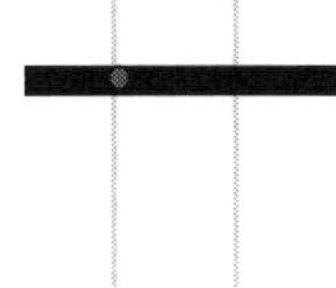		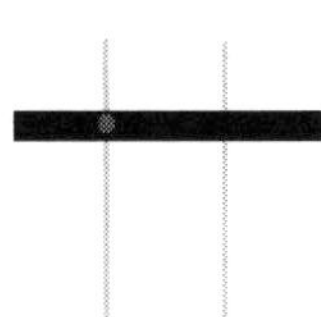		Wednesday
50	**13**	**04**	**53**	**34**	
					Thursday
55	**03**	**42**	**14**	**22**	
					Friday

WEEK 6 – LESSON 6 – MIND MATH – Numbers 6 and 7

LESSON 6 –PRACTICE WORK

Use Abacus

DAY 1 – MONDAY

TIME: ________min __________sec Accuracy ______/16

6:1

1	2	3	4	5	6	7	8
53	85	55	04	65	87	28	35
21	14	03	25	24	- 26	20	53
- 14	- 95	01	50	- 02	32	- 03	10
30	50	- 56	- 02	- 71	- 50	01	- 18

6:2

1	2	3	4	5	6	7	8
72	43	31	01	28	37	50	56
22	55	18	12	20	- 25	15	22
- 30	01	- 15	11	- 35	20	- 50	- 11
15	- 99	- 14	- 24	50	01	33	- 62

DAY 2 – TUESDAY

TIME: ________min __________sec Accuracy ______/16

6:3

1	2	3	4	5	6	7	8
11	13	24	56	33	93	20	84
20	20	25	23	16	- 80	23	- 50
57	- 22	50	- 59	50	21	51	05
10	82	- 18	68	- 47	50	- 94	- 37

6:4

1	2	3	4	5	6	7	8
99	49	85	50	26	77	51	89
- 24	50	14	16	23	- 20	45	- 26
- 50	- 16	- 70	13	- 48	22	- 50	35
01	- 11	- 10	- 20	56	- 79	03	- 78

DAY 3 – WEDNESDAY

TIME: __________min ____________sec Accuracy _______/16

1	2	3	4	5	6	7	8
51	52	52	61	10	31	72	94
11	05	02	12	71	51	- 61	- 60
11	41	45	15	- 61	- 21	82	- 12
11	- 76	- 51	- 12	22	18	01	76

6:5

1	2	3	4	5	6	7	8
56	22	51	23	62	25	54	87
23	51	42	- 02	25	22	- 03	11
- 62	25	- 73	75	- 31	- 46	15	- 75
71	- 87	65	- 85	41	15	21	51

6:6

DAY 4 – THURSDAY

TIME: __________min ____________sec Accuracy _______/16

1	2	3	4	5	6	7	8
83	33	39	83	53	27	29	42
- 52	11	- 11	- 51	- 01	- 21	- 15	- 32
65	55	- 16	05	45	93	80	23
01	- 63	- 11	- 32	- 87	- 65	- 64	- 33

6:7

1	2	3	4	5	6	7	8
53	94	73	50	36	50	32	34
- 02	- 53	- 62	23	- 21	10	11	15
45	05	25	- 11	74	01	56	- 28
- 76	- 31	61	- 61	- 85	38	- 19	76

6:8

DAY 5 – FRIDAY

TIME: __________min ____________sec Accuracy ______/16

1	2	3	4	5	6	7	8
53	68	86	24	66	37	12	62
16	- 02	- 35	55	22	- 22	55	11
- 11	- 15	17	- 12	01	14	11	11
- 55	27	- 55	- 05	- 52	- 27	- 27	- 72

6:9

1	2	3	4	5	6	7	8
93	27	99	54	22	12	72	21
- 52	51	- 15	25	72	11	15	21
- 20	- 12	- 30	- 18	- 81	51	12	01
58	- 56	- 54	- 60	- 01	25	- 60	51

6:10

LESSON 6 – SKILL BUILDING EXERCISE

Draw beads to show numbers given.

85 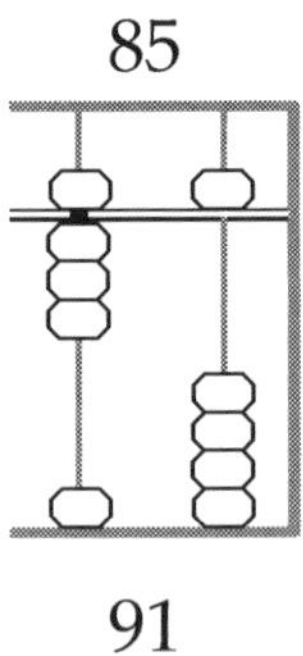

64 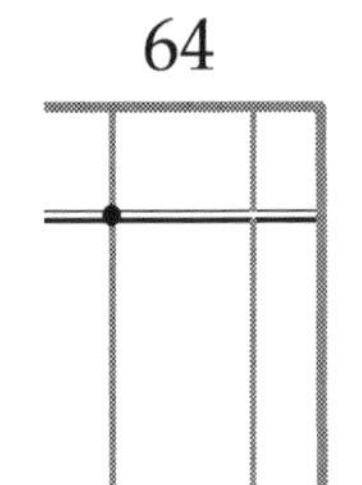

43 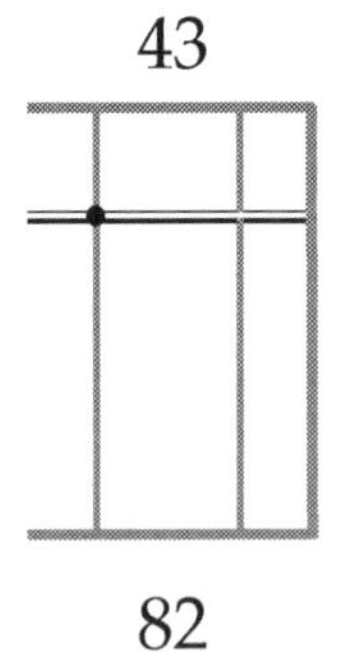

39 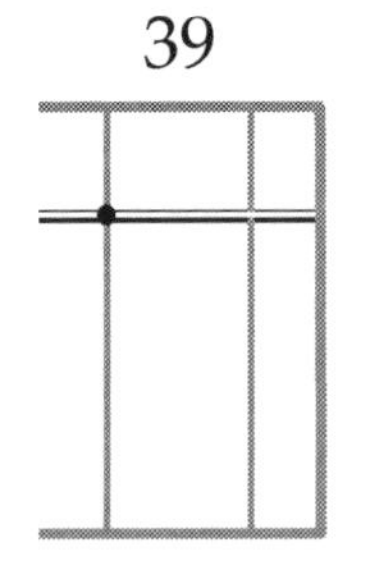

62 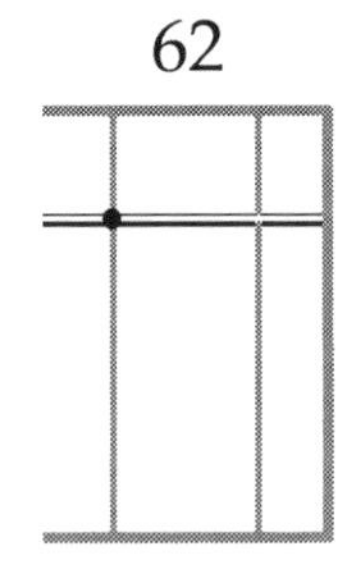

Tuesday

91

15 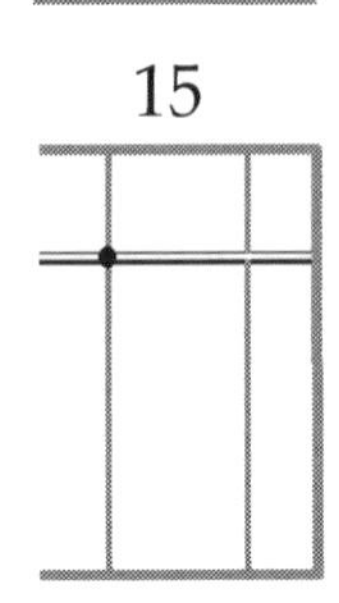

82 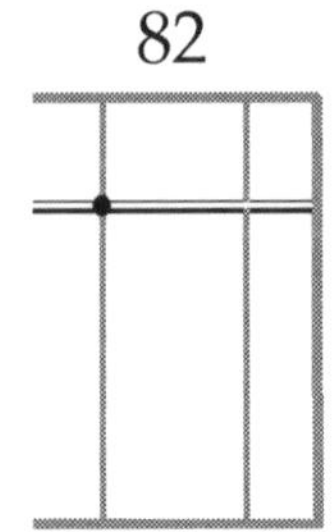

26 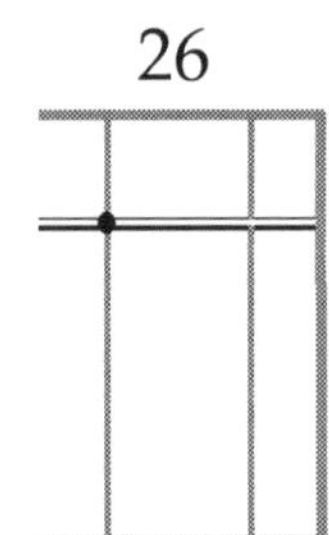

59 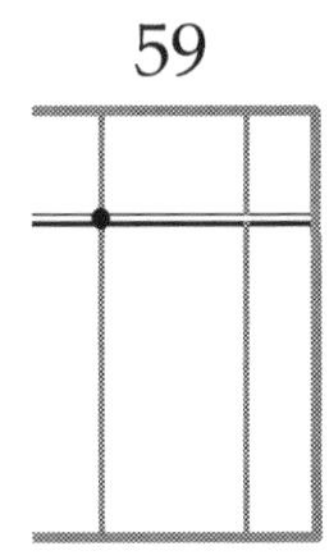

Thursday

LESSON 6 – MIND MATH PRACTICE

Visualize 

DAY 1 – MONDAY

Accuracy ______/40 ☆

1	2	3	4	5	6	7	8	9	10
02	01	01	07	06	05	02	07	16	12
02	01	05	- 01	01	02	02	- 02	01	05
- 03	05	- 01	01	- 02	- 01	- 01	- 05	- 05	- 01

611 —

1	2	3	4	5	6	7	8	9	10
10	20	50	10	70	70	20	70	70	60
50	50	10	50	- 20	- 10	50	- 60	- 50	- 10
- 50	- 10	- 10	10	10	- 50	- 60	50	- 10	20

6:12 —

1	2	3	4	5	6	7	8	9	10
62	72	52	61	11	72	33	64	64	54
02	- 02	02	- 10	11	01	- 11	- 02	- 02	- 50
- 03	05	- 50	05	02	- 50	55	- 02	05	10

6:13 —

1	2	3	4	5	6	7	8	9	10
55	06	02	10	44	40	31	74	64	24
01	11	55	10	- 21	- 30	10	- 20	- 03	- 03
- 50	- 05	- 01	55	- 20	50	- 40	- 02	05	11

6:14 —

DAY 2 – TUESDAY

Accuracy ______/10 ☆

1	2	3	4	5	6	7	8	9	10
05	06	27	52	21	50	70	51	51	27
02	01	- 22	02	01	20	- 60	20	05	- 02
- 06	- 07	01	- 54	05	- 50	10	- 70	- 01	21

6:15 —

DAY 3 – WEDNESDAY

Accuracy ______/10 ☆

1	2	3	4	5	6	7	8	9	10
70	60	70	60	65	55	60	14	50	60
- 50	10	- 20	- 10	- 10	- 05	06	- 03	20	- 50
20	- 50	10	- 50	20	02	- 01	15	- 50	30

6:16

DAY 4 – THURSDAY

Accuracy ______/10 ☆

1	2	3	4	5	6	7	8	9	10
05	30	64	73	55	22	44	44	54	77
40	- 20	- 50	01	22	11	- 03	- 30	- 02	- 11
- 10	15	- 01	- 04	- 20	- 33	05	- 10	05	- 55

6:17

DAY 5 – FRIDAY

Accuracy ______/20 ☆

1	2	3	4	5	6	7	8	9	10
71	44	07	60	71	65	77	52	61	11
02	- 12	70	15	02	02	- 12	21	10	11
- 01	- 12	- 05	- 50	01	- 06	- 05	- 03	- 20	50

6:18

1	2	3	4	5	6	7	8	9	10
05	02	30	72	33	72	33	54	74	54
22	05	06	- 10	- 20	01	- 11	- 02	- 02	- 50
- 11	20	- 30	- 11	50	- 50	55	- 52	05	60

6:19

LESSON 6 – SKILL BUILDING

Visualize the numbers on the beam in your mind and draw it to represent the numbers given.

31 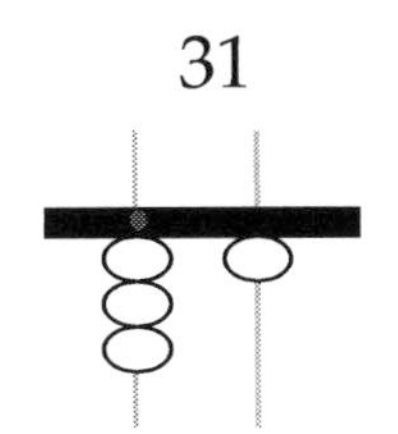	55 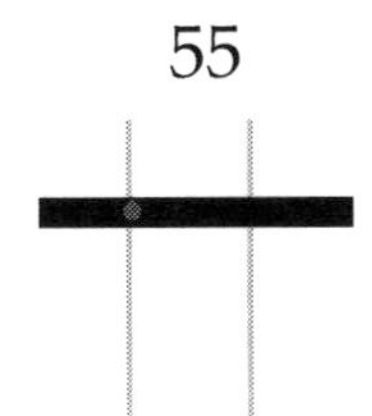	15 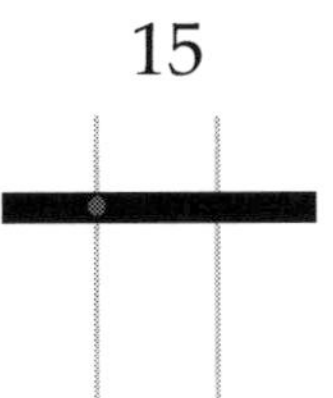	46 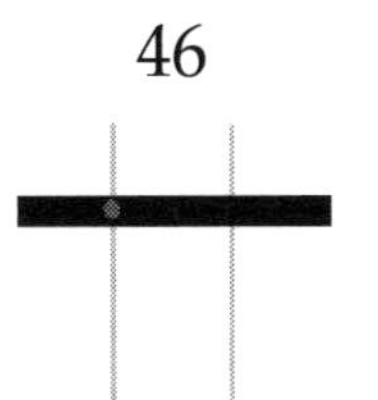	57 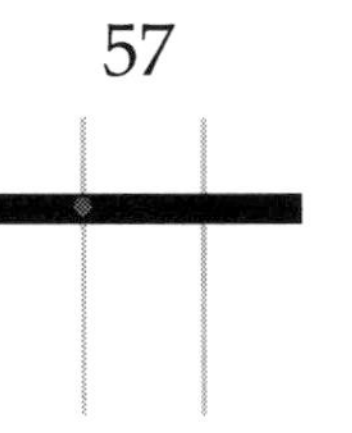	Monday
63 	22 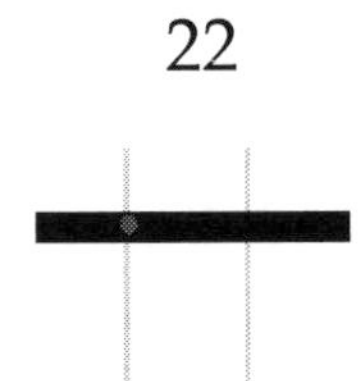	14	33 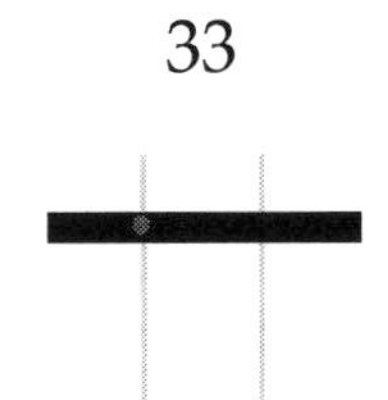	47 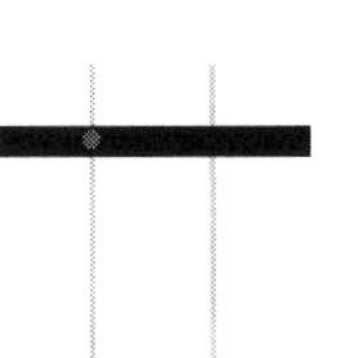	Tuesday
45	06	62	17	36	Wednesday
71	36	74	27	13	Thursday
42	53	75	65	07	Friday

WEEK 7 – LESSON 7 – MIND MATH – Numbers 8 and 9

Use Abacus

LESSON 7 –PRACTICE WORK

DAY 1 – MONDAY

TIME: __________min ____________sec Accuracy ______/16

1	2	3	4	5	6	7	8
01	42	52	12	57	88	42	32
51	01	01	02	21	- 32	- 31	- 21
41	01	25	10	- 66	43	02	- 01
01	50	10	55	52	- 69	11	31

7:1

1	2	3	4	5	6	7	8
53	85	52	36	14	72	64	84
40	01	42	- 21	15	22	- 53	10
- 82	12	- 63	64	- 09	- 84	25	- 33
58	- 71	55	- 50	12	75	12	25

7:2

DAY 2 – TUESDAY

TIME: __________min ____________sec Accuracy ______/16

1	2	3	4	5	6	7	8
15	44	74	73	97	66	45	41
12	- 01	- 12	- 02	- 67	- 06	53	52
11	55	25	25	58	07	- 61	- 91
51	- 71	- 15	- 86	- 66	- 66	- 21	07

7:3

1	2	3	4	5	6	7	8
15	12	83	43	93	23	44	79
21	82	- 12	51	- 31	- 11	- 31	- 23
53	05	25	- 62	- 11	77	76	- 51
- 70	- 71	- 75	50	- 51	- 81	- 61	64

7:4

DAY 3 – WEDNESDAY

TIME: ________min ________sec Accuracy ______/16

1	2	3	4	5	6	7	8
53	41	62	22	52	11	32	93
35	02	32	15	11	52	51	- 62
- 77	- 33	- 71	11	11	21	01	56
86	09	- 11	50	- 23	15	- 73	- 85

7:5

1	2	3	4	5	6	7	8
51	77	22	42	32	65	42	44
25	22	57	51	52	34	52	- 13
11	- 86	- 71	01	- 03	- 81	- 74	56
- 36	15	50	- 60	- 81	71	69	- 72

7:6

DAY 4 – THURSDAY

TIME: ________min ________sec Accuracy ______/16

1	2	3	4	5	6	7	8
56	71	21	32	44	68	44	83
42	- 20	52	52	- 03	- 12	50	- 11
- 63	41	- 11	- 74	52	- 55	- 70	- 01
51	- 80	36	05	- 11	77	60	- 01

7:7

1	2	3	4	5	6	7	8
74	83	12	84	83	53	35	74
- 11	- 31	32	- 23	- 11	25	11	- 23
25	05	- 44	05	- 21	- 56	52	45
- 08	- 01	57	31	45	25	- 86	- 80

7:8

DAY 5 – FRIDAY

TIME: __________min __________sec Accuracy ______/16 ☆

1	2	3	4	5	6	7	8
79	51	35	94	23	72	33	55
- 13	28	54	- 41	51	- 21	15	12
21	- 65	- 25	30	- 72	- 01	- 20	- 01
- 55	55	- 64	05	85	42	71	31

7:9

1	2	3	4	5	6	7	8
68	22	92	11	25	91	66	15
20	51	05	83	53	03	13	73
- 37	- 53	- 67	- 44	- 22	05	- 29	- 15
45	- 20	51	06	40	- 51	- 50	- 13

7:10

LESSON 7 – SKILL BUILDING EXERCISE

Draw beads to show numbers given.

85

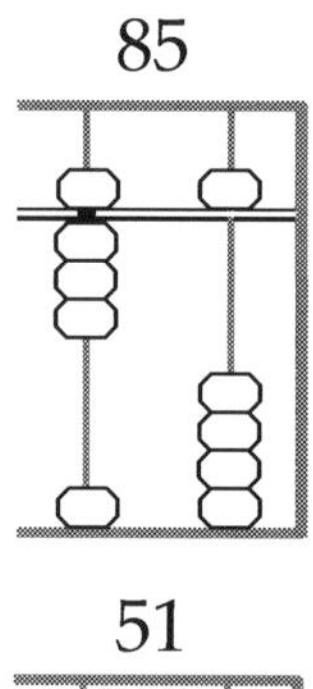

65

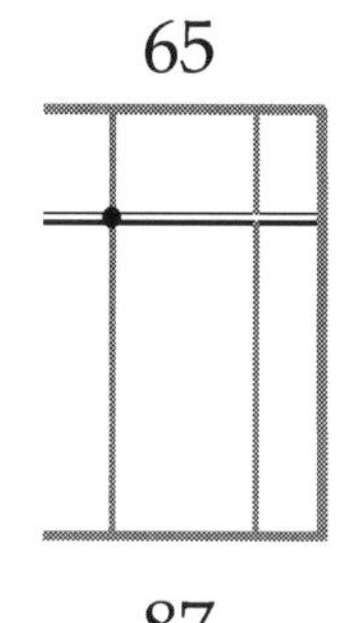

44

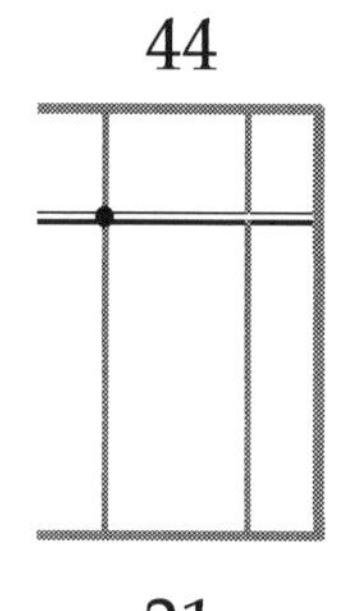

38

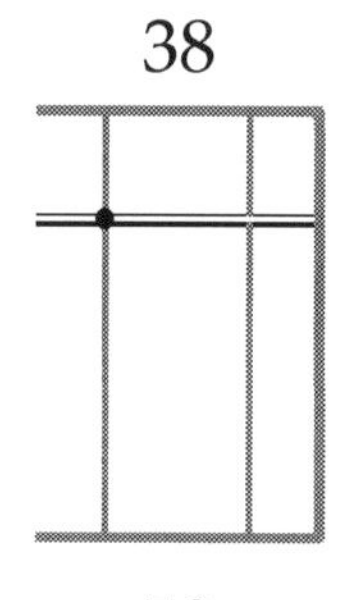

63

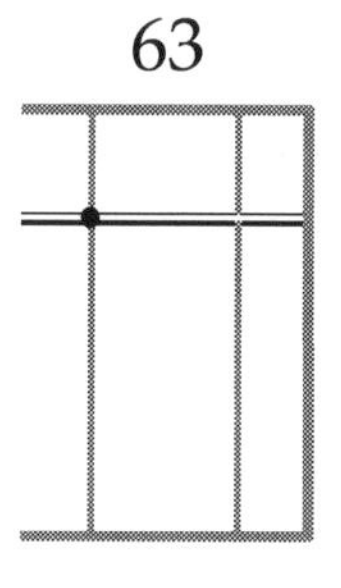

Tuesday

51

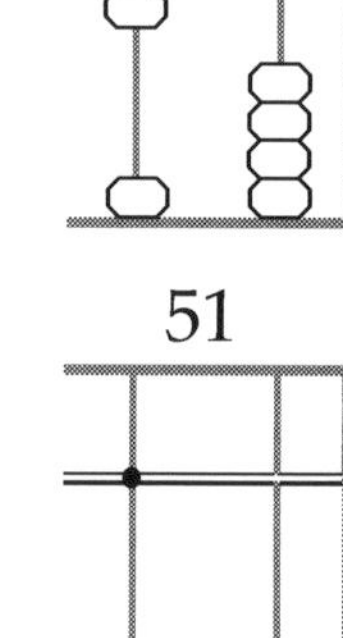

87

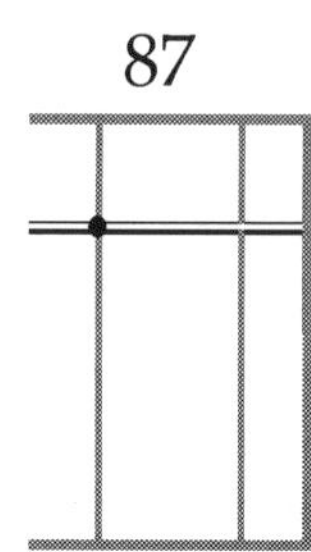

21

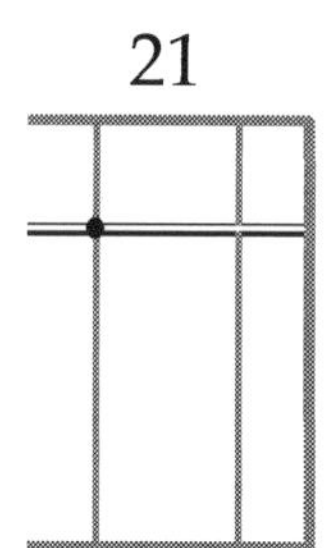

50

08

Thursday

LESSON 7 – MIND MATH PRACTICE

DAY 1 – MONDAY

Accuracy ______/40 ☆

7:11

1	2	3	4	5	6	7	8	9	10
05	02	02	07	08	06	07	09	07	03
01	01	05	02	- 02	02	02	- 02	02	05
01	05	02	- 01	- 01	- 01	- 01	- 02	- 05	- 06

7:12

1	2	3	4	5	6	7	8	9	10
70	80	90	90	70	70	30	75	77	65
- 20	10	- 10	- 10	- 10	10	50	- 60	- 56	- 10
- 50	- 50	10	- 20	30	- 80	10	30	70	40

7:13

1	2	3	4	5	6	7	8	9	10
10	20	50	20	40	62	73	52	63	12
50	20	30	10	50	02	- 02	02	- 11	70
20	- 10	- 20	50	- 10	- 01	05	05	40	- 02

7:14

1	2	3	4	5	6	7	8	9	10
72	44	74	64	64	05	12	24	75	83
20	- 11	- 03	- 12	- 13	20	05	05	- 10	- 20
07	55	08	25	08	60	22	- 11	- 10	30

DAY 2 – TUESDAY

Accuracy ______/10 ☆

7:15

1	2	3	4	5	6	7	8	9	10
25	12	53	94	64	75	80	94	23	13
02	32	05	- 03	05	20	05	- 02	55	51
01	05	- 02	05	- 01	- 10	10	- 02	01	- 02

DAY 3 – WEDNESDAY

Accuracy ______/10 ☆

1	2	3	4	5	6	7	8	9	10
01	04	03	20	90	38	51	84	93	74
05	05	06	70	01	- 20	03	- 03	- 11	- 03
03	- 01	- 02	05	05	50	05	- 01	- 02	05

7:16

DAY 4 – THURSDAY

Accuracy ______/10 ☆

1	2	3	4	5	6	7	8	9	10
07	03	63	50	70	40	31	79	84	64
- 05	05	- 01	30	- 50	- 10	10	- 02	05	- 03
06	11	05	- 70	20	50	50	20	- 09	11

7:17

DAY 5 – FRIDAY

Accuracy ______/20 ☆

1	2	3	4	5	6	7	8	9	10
15	36	54	43	55	22	84	99	94	64
12	- 20	05	- 01	20	51	- 03	- 11	- 42	- 12
01	03	- 01	50	03	- 03	10	- 03	- 50	05

7:18

1	2	3	4	5	6	7	8	9	10
62	43	62	33	72	60	91	85	67	44
02	- 02	- 12	- 22	- 61	10	02	04	02	55
20	08	08	80	07	- 20	- 80	- 80	- 50	- 10

7:19

LESSON 7 – SKILL BUILDING

Visualize the numbers on the beam in your mind and draw it to represent the numbers given.

78 57 18 49 76

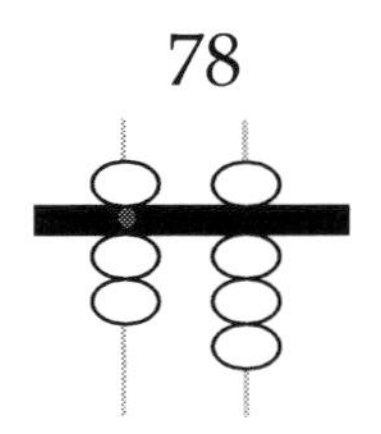
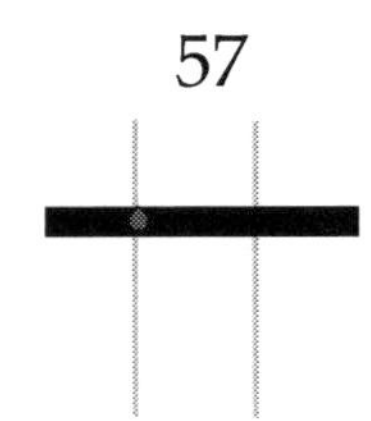
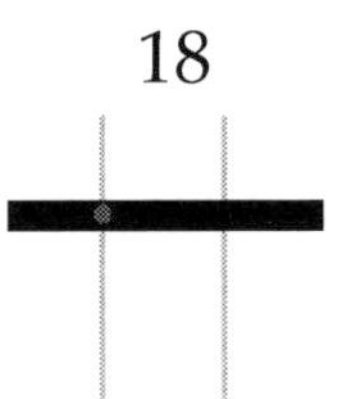
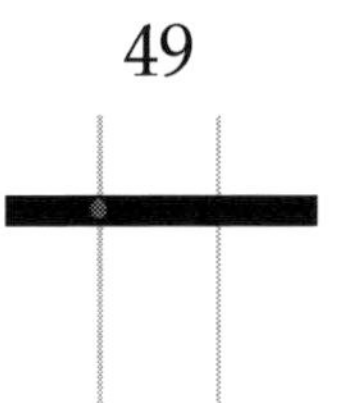
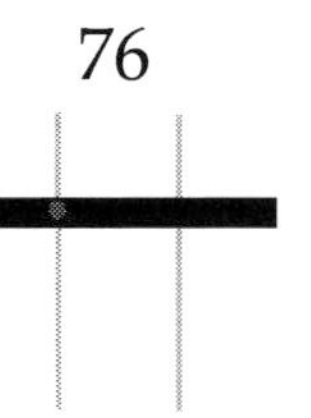

Monday

94 23 04 35 41

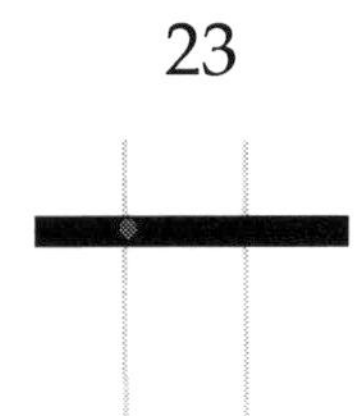
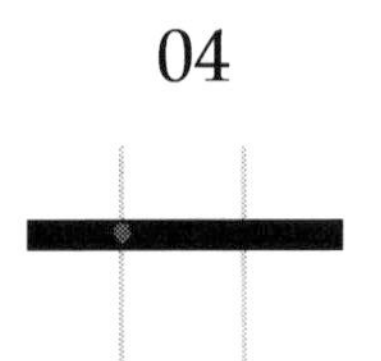
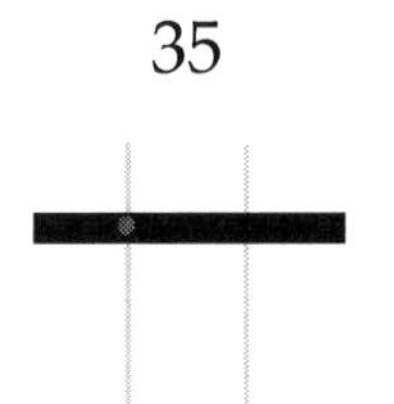
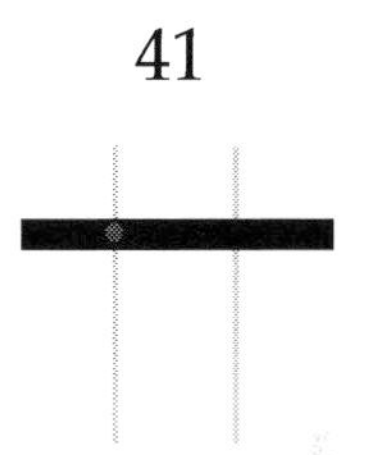

Tuesday

65 11 19 27 72

Wednesday

81 93 74 29 58

Thursday

96 77 89 67 25

Friday

WEEK 8 – LESSON 8 – SMALL FRIEND COMBINATIONS

SMALL FRIEND – COMBINATIONS OF 5

1 and 4 are friends of 5

1 + 4 = 5   4 + 1 = 5

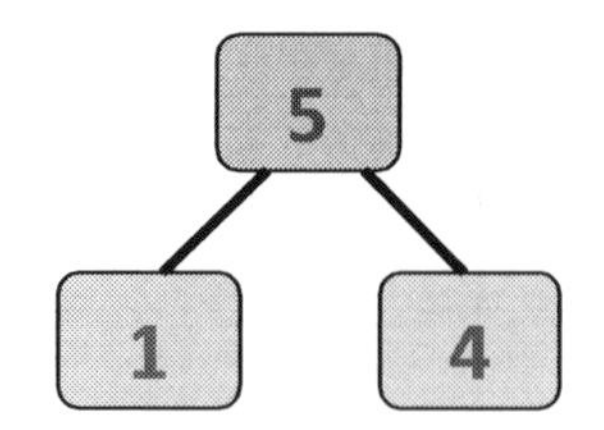

2 and 3 are friends of 5

2 + 3 = 5   3 + 2 = 5

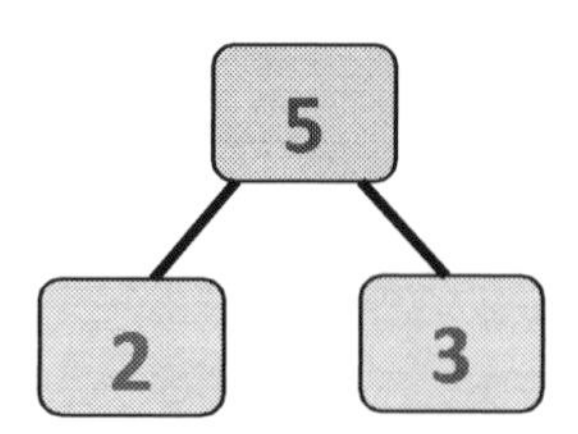

NUMBER SENTENCE FOR COMBINATIONS OF 5

Addition Facts	Subtraction Facts
1 + 4 = 5 	5 – 1 = 4 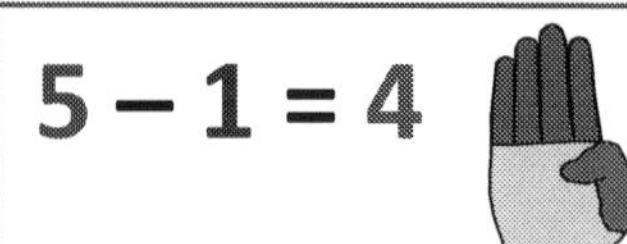
4 + 1 = 5 	5 – 4 = 1 
2 + 3 = 5 	5 – 2 = 3 
3 + 2 = 5 	5 – 3 = 2 

1 and 4 are small friends, because together they make 5.
2 and 3 are small friends, because together they make 5.

LESSON 8 – ACTIVITY – 1

DRAW A PICTURE (of anything) TO SHOW THE NUMBER SENTENCES

Example: 4 + 1 = 5

Example: 5 – 3 = 2

DAY 1 - MONDAY

1 + 4 = 5 4 + 1 = 5 2 + 3 = 5 3 + 2 = 5

+ =

5 – 4 = 1 5 – 1 = 4 5 – 2 = 3 5 – 3 = 2

=

DAY 2 - TUESDAY

1 + 4 = 5 4 + 1 = 5 2 + 3 = 5 3 + 2 = 5

5 – 4 = 1 5 – 1 = 4 5 – 2 = 3 5 – 3 = 2

DAY 3 - WEDNESDAY

1 + 4 = 5 4 + 1 = 5 2 + 3 = 5 3 + 2 = 5

5 – 4 = 1 5 – 1 = 4 5 – 2 = 3 5 – 3 = 2

DAY 4 - THURSDAY

$1 + 4 = 5$ $4 + 1 = 5$ $2 + 3 = 5$ $3 + 2 = 5$

$5 - 4 = 1$ $5 - 1 = 4$ $5 - 2 = 3$ $5 - 3 = 2$

DAY 5 - FRIDAY

$1 + 4 = 5$ $4 + 1 = 5$ $2 + 3 = 5$ $3 + 2 = 5$

$5 - 4 = 1$ $5 - 1 = 4$ $5 - 2 = 3$ $5 - 3 = 2$

DAY 6 - SATURDAY

$1 + 4 = 5$ $4 + 1 = 5$ $2 + 3 = 5$ $3 + 2 = 5$

$5 - 4 = 1$ $5 - 1 = 4$ $5 - 2 = 3$ $5 - 3 = 2$

LESSON 8 – ACTIVITY – 2

Color the circle if the number sentence is true. Cross the circle if the number sentence is wrong.

Example: ● 4 + 1 = 5 ⊗ 2 + 4 = 5 ● 5 – 2 = 3

DAY 1	DAY 2	DAY 3
○ 2 + 3 = 5	○ 5 – 4 = 1	○ 5 – 3 = 2
○ 1 + 4 = 1	○ 5 – 2 = 1	○ 5 + 4 = 1
○ 2 + 2 = 5	○ 1 + 4 = 5	○ 1 + 4 = 5
○ 4 + 1 = 5	○ 3 + 2 = 5	○ 5 – 2 = 3
○ 5 – 4 = 1	○ 5 – 3 = 4	○ 3 + 2 = 5
○ 1 + 4 = 5	○ 5 – 2 = 3	○ 1 – 5 = 2
○ 5 – 2 = 3	○ 5 + 5 = 3	○ 5 – 1 = 4
○ 3 + 2 = 5	○ 5 – 3 = 2	○ 4 + 1 = 5
○ 4 + 1 = 3	○ 4 + 1 = 5	○ 3 + 2 = 4

DAY 4	DAY 5	DAY 6
○ 4 + 1 = 5	○ 4 + 3 = 2	○ 5 – 3 = 3
○ 5 – 3 = 2	○ 1 + 4 = 5	○ 5 – 1 = 4
○ 5 – 2 = 1	○ 5 – 3 = 4	○ 5 – 3 = 2
○ 1 + 4 = 5	○ 2 + 3 = 5	○ 6 + 5 = 5
○ 5 – 2 = 3	○ 5 – 2 = 3	○ 5 – 1 = 4
○ 5 + 5 = 3	○ 4 + 1 = 5	○ 2 + 3 = 5
○ 3 + 2 = 5	○ 5 + 5 = 3	○ 5 – 4 = 1
○ 5 – 4 = 1	○ 5 – 4 = 1	○ 2 + 3 = 5
○ 5 – 3 = 4	○ 5 – 2 = 5	○ 5 – 1 = 5

LESSON 8 – PRACTICE WORK

Use Abacus

DAY 1 – MONDAY

TIME: __________min __________sec Accuracy ______/16

1	2	3	4	5	6	7	8
11	10	24	10	33	12	22	61
21	27	25	20	11	30	22	36
55	- 16	50	59	50	55	50	- 41
- 62	27	- 71	10	- 40	- 11	- 31	22

8:1

1	2	3	4	5	6	7	8
21	38	17	14	20	12	55	62
11	10	21	25	28	56	32	27
55	50	- 18	- 12	51	- 15	- 25	- 35
10	- 27	25	21	- 30	41	17	- 52

8:2

© SAI Speed Math Academy, USA

DAY 2 – TUESDAY

TIME: __________min __________sec Accuracy ______/16

1	2	3	4	5	6	7	8
56	36	12	72	64	83	41	98
12	- 21	15	07	- 53	- 12	53	- 53
- 57	54	- 01	- 64	25	25	05	04
35	- 11	12	52	11	- 81	- 87	- 35

8:3

© SAI Speed Math Academy, USA

1	2	3	4	5	6	7	8
28	43	11	21	92	25	37	45
- 11	- 31	35	62	- 41	12	11	51
70	85	52	15	06	51	51	- 36
- 82	- 61	- 25	- 21	- 51	- 15	- 40	17

8:4

© SAI Speed Math Academy, USA

DAY 3 – WEDNESDAY

TIME: ________min __________sec Accuracy ______/16

1	2	3	4	5	6	7	8	8:5
78 - 17 26 - 50	33 51 - 82 - 02	35 52 10 - 26	29 20 - 39 50	41 52 - 11 - 11	38 11 - 32 50	50 13 15 20	21 21 51 - 93	

1	2	3	4	5	6	7	8	8:6
39 - 28 75 03	54 05 - 56 91	24 50 - 71 55	42 51 05 - 13	32 52 - 03 - 81	65 22 12 - 45	42 52 - 74 25	87 - 11 - 75 18	

DAY 4 – THURSDAY

TIME: ________min __________sec Accuracy ______/16

1	2	3	4	5	6	7	8	8:7
15 12 11 51	44 - 01 55 - 71	84 - 12 25 - 12	73 - 02 25 - 86	97 - 66 58 - 77	66 21 - 70 82	45 53 - 61 - 37	46 03 - 39 85	

1	2	3	4	5	6	7	8	8:8
35 51 - 15 08	94 - 82 55 - 12	67 32 - 44 31	84 - 23 05 31	83 - 11 07 - 19	53 25 - 56 25	35 11 52 - 95	49 - 18 52 - 81	

DAY 5 – FRIDAY

TIME: __________min ____________sec Accuracy _______/16 ☆

1	2	3	4	5	6	7	8
46	39	84	44	57	43	97	27
50	- 06	15	55	32	- 22	- 55	51
- 11	15	- 17	- 67	10	68	- 11	11
- 55	- 27	- 61	05	- 58	- 27	18	- 72

8:9

1	2	3	4	5	6	7	8
67	45	98	33	82	76	49	15
- 52	51	- 53	65	- 71	22	- 15	22
20	- 65	- 30	- 18	86	- 56	- 12	11
14	56	12	- 70	- 90	07	- 22	- 35

8:10

LESSON 8 – SKILL BUILDING EXERCISE

Draw beads to show numbers given.

15

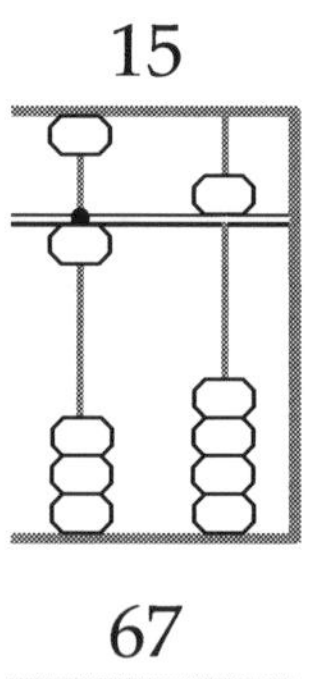

28

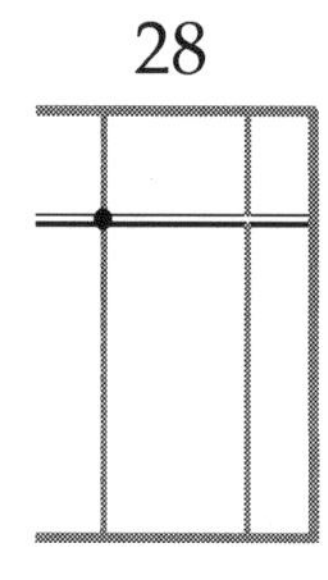

32

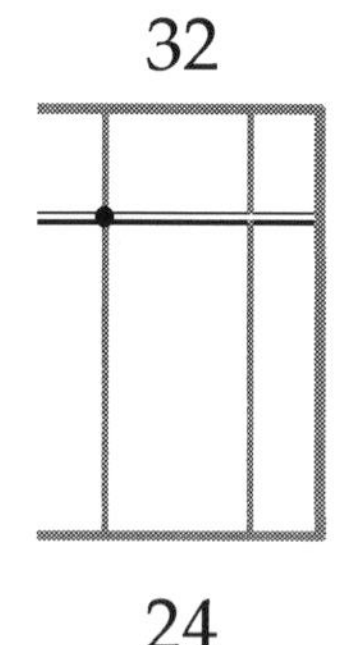

52

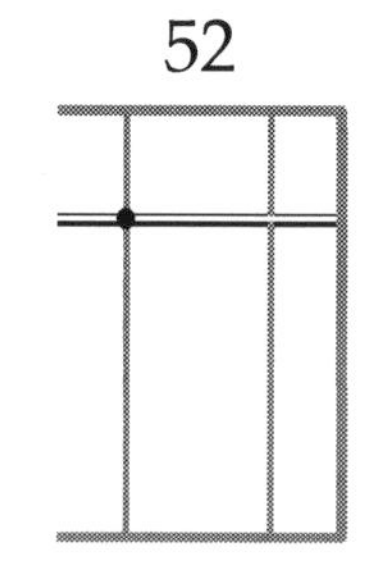

88

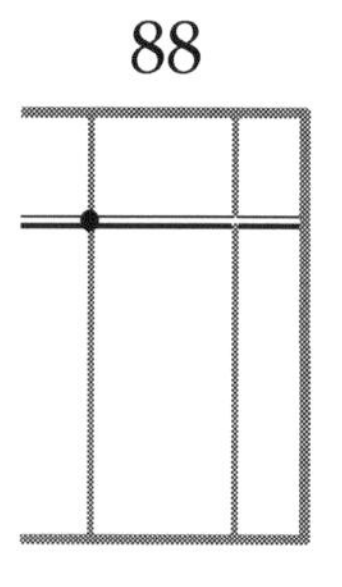

67

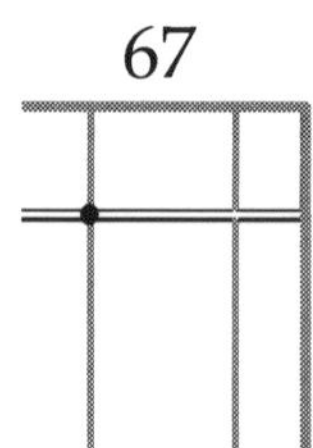

99

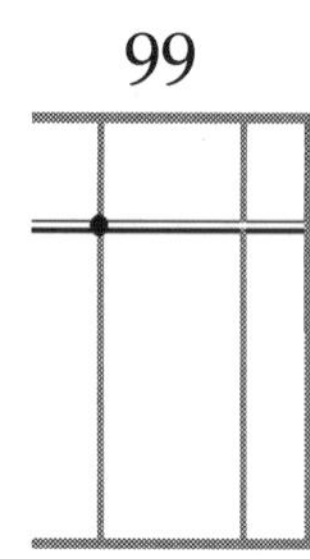

24

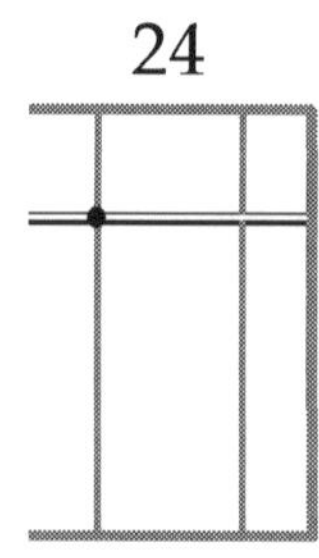

54

48

Thursday

LESSON 8 – MIND MATH PRACTICE

DAY 1 – MONDAY

Accuracy ______/10 ☆

1	2	3	4	5	6	7	8	9	10
22	33	19	53	51	66	44	44	64	93
50	- 02	30	- 01	08	33	- 31	05	- 12	06
- 01	50	- 44	25	- 07	- 08	05	- 08	05	- 49

8:11

DAY 2 – TUESDAY

Accuracy ______/10 ☆

1	2	3	4	5	6	7	8	9	10
					27	29	42	53	94
12	28	57	22	49	- 21	- 15	- 31	- 02	- 53
52	21	30	11	- 20	93	80	23	40	05
10	- 09	- 26	50	- 25	- 55	- 40	- 33	- 91	- 31

8:12

DAY 3 – WEDNESDAY

Accuracy ______/10 ☆

1	2	3	4	5	6	7	8	9	10
					22	65	84	28	63
40	49	41	08	52	- 01	32	- 73	- 02	20
51	- 02	52	- 01	01	- 10	- 51	10	- 11	- 72
- 20	- 02	- 30	30	06	70	03	07	24	50

8:13

DAY 4 – THURSDAY

Accuracy ______/10 ☆

1	2	3	4	5	6	7	8	9	10
60	11	88	52	11	73	50	36	50	32
02	50	- 21	- 01	35	- 62	23	- 21	10	11
02	- 10	- 01	25	- 01	25	- 11	70	01	06
15	- 51	- 65	- 11	- 10	60	- 61	- 85	38	- 40

8:14

DAY 5 – FRIDAY Accuracy ______/20 ☆

1	2	3	4	5	6	7	8	9	10
16	04	44	94	64	13	51	47	33	74
13	- 03	- 21	- 22	- 12	- 02	06	- 10	01	- 23
- 09	07	15	- 11	06	01	- 50	- 12	15	40

8:15

1	2	3	4	5	6	7	8	9	10
41	02	90	33	41	88	42	32	53	85
02	02	- 10	- 20	- 20	- 30	- 31	- 21	40	01
06	- 03	- 30	05	- 20	40	02	- 01	- 80	02
- 07	78	05	- 18	55	- 90	11	80	05	- 71

8:16

LESSON 8 – SKILL BUILDING

Visualize the numbers on the beam in your mind and draw it to represent the numbers given.

78 61 05 73 85

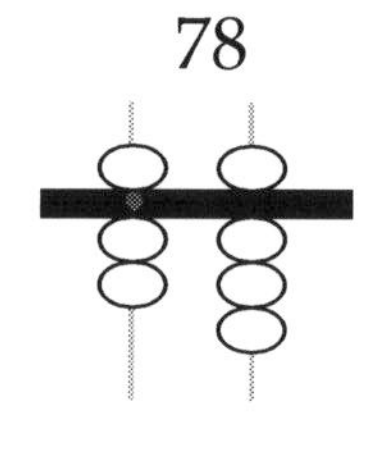
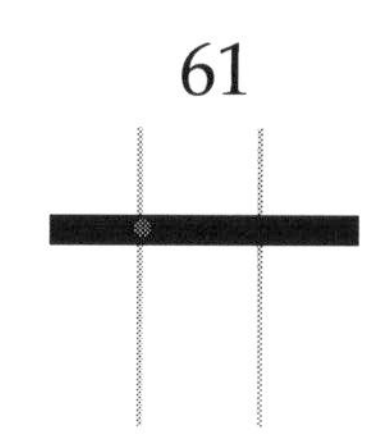

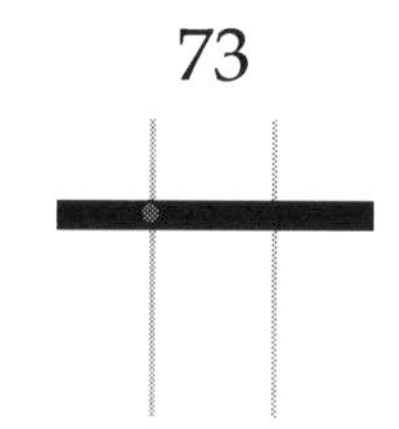
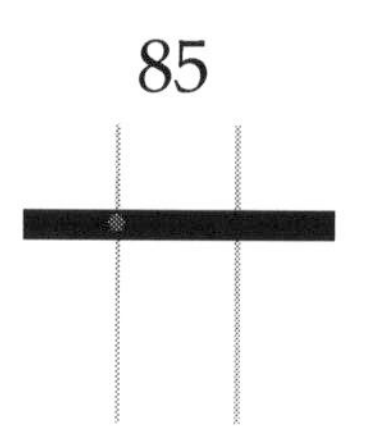

69 17 82 56 35

Wednesday

WEEK 9 – LESSON 9 – INTRODUCING +1 CONCEPT

LESSON 9 – PRACTICE WORK

+1 = + 5 – 4

DAY 1 – MONDAY

Accuracy ______/32 ☆

Use Abacus

1	2	3	4	5	6	7	8
10	24	20	43	34	23	25	35
10	21	13	- 03	- 13	01	24	03
24	14	11	15	23	21	- 05	11
01	- 05	11	- 05	11	01	11	- 04

9:1

1	2	3	4	5	6	7	8
14	05	25	12	24	23	33	24
21	04	24	32	21	11	11	11
14	- 02	- 35	11	50	11	51	14
- 39	- 01	11	23	- 10	12	- 40	- 26

9:2

1	2	3	4	5	6	7	8
47	52	94	13	23	14	20	49
11	12	01	11	11	50	24	10
- 06	31	- 50	21	11	01	- 30	- 05
- 02	- 95	10	13	12	- 15	51	01

9:3

1	2	3	4	5	6	7	8
49	34	80	53	44	43	54	27
- 02	11	10	41	11	11	01	- 02
10	50	04	- 90	44	11	- 50	23
- 56	- 35	01	71	- 75	34	94	11

9:4

DAY 2 – TUESDAY

TIME: __________min ____________sec Accuracy _______/16

1	2	3	4	5	6	7	8
13	22	34	21	32	17	20	44
31	22	- 32	11	12	- 15	20	10
51	51	52	52	51	22	50	21
- 50	- 50	- 50	11	- 20	51	- 30	- 50

9:5

1	2	3	4	5	6	7	8
42	44	33	50	24	34	54	99
12	51	11	14	21	- 22	01	- 56
41	- 10	11	11	- 40	31	- 50	15
04	- 10	44	20	50	10	60	- 58

9:6

DAY 3 – WEDNESDAY

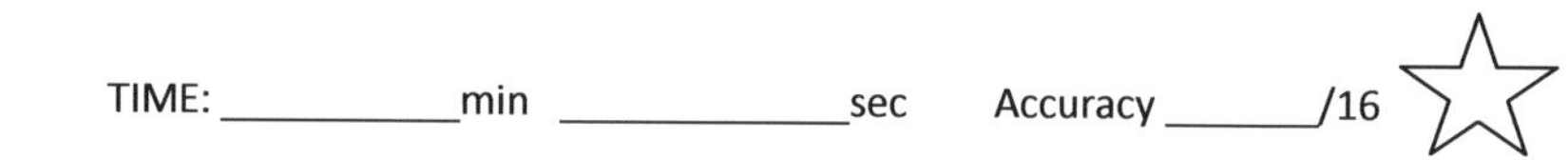

TIME: __________min ____________sec Accuracy _______/16

1	2	3	4	5	6	7	8
14	73	35	24	64	34	04	43
11	- 21	51	21	31	11	11	11
20	01	- 66	- 30	01	- 10	- 10	10
51	40	05	01	- 10	- 15	04	- 52

9:7

1	2	3	4	5	6	7	8
44	50	13	23	04	20	41	41
01	05	11	21	50	24	52	18
11	- 50	21	51	01	- 30	- 20	- 55
01	44	51	01	- 50	01	- 13	61

9:8

DAY 4 – THURSDAY

TIME: __________min __________sec Accuracy ______/16 ☆

1	2	3	4	5	6	7	8
44	25	47	31	42	40	02	33
51	21	10	11	12	05	42	11
- 20	53	- 06	12	15	10	- 03	- 40
- 10	- 59	- 01	15	10	- 05	10	01

9:9

1	2	3	4	5	6	7	8
73	43	74	22	37	24	32	30
- 12	11	01	22	- 21	51	02	14
17	05	- 60	- 40	71	14	61	11
- 76	- 53	- 10	01	- 67	- 58	- 85	41

9:10

DAY 5 – FRIDAY

TIME: __________min __________sec Accuracy ______/16 ☆

1	2	3	4	5	6	7	8
14	88	23	76	57	84	32	25
81	- 53	- 11	- 15	42	11	55	11
- 55	12	32	33	- 56	- 50	11	11
17	11	11	01	15	13	- 97	12

9:11

1	2	3	4	5	6	7	8
40	44	34	34	51	14	36	54
14	- 01	51	11	13	21	- 21	01
20	11	11	01	11	- 35	30	- 55
- 51	30	- 10	52	20	50	11	83

9:12

LESSON 9 – MIND MATH PRACTICE

Visualize

DAY 1 – MONDAY

Accuracy ______/10 ☆

1	2	3	4	5	6	7	8	9	10
15	04	43	11	24	34	04	23	44	13
11	21	01	33	01	01	01	01	15	01
- 06	04	- 22	01	- 10	- 05	10	01	- 01	01

9:13

DAY 2 – TUESDAY

Accuracy ______/10 ☆

1	2	3	4	5	6	7	8	9	10
					43	50	24	14	10
12	12	04	14	54	01	10	01	01	04
12	10	20	01	- 01	11	- 60	21	04	11
11	05	01	01	- 01	- 05	11	10	- 09	- 05

9:14

DAY 3 – WEDNESDAY

Accuracy ______/10 ☆

1	2	3	4	5	6	7	8	9	10
					15	04	24	40	84
34	04	33	44	24	21	- 03	01	10	- 10
01	01	10	01	01	10	02	02	- 50	01
10	- 05	- 20	- 10	01	10	- 01	01	70	- 20

9:15

DAY 4 – THURSDAY

Accuracy ______/10 ☆

1	2	3	4	5	6	7	8	9	10
					14	20	04	44	05
41	43	22	47	21	- 11	20	51	11	04
10	11	22	10	01	01	10	01	- 50	- 02
10	- 52	01	- 02	50	- 04	30	- 50	- 05	- 02

9:16

DAY 5 – FRIDAY

1	2	3	4	5	6	7	8	9	10	9:17
					94	43	51	80	64	
40	44	45	41	32	01	- 12	03	- 70	11	
02	11	- 10	10	10	- 55	03	41	60	- 50	
10	- 50	- 20	07	17	10	01	04	- 20	- 25	

1	2	3	4	5	6	7	8	9	10	9:18
44	22	34	44	34	49	28	40	76	34	
01	12	51	11	51	- 35	- 01	03	- 51	11	
10	01	01	- 50	- 10	80	- 05	01	- 20	14	
- 55	- 30	- 61	73	- 10	01	21	11	60	- 08	

SHAPE PUZZLE

Find out what each shape is worth.

Suggestion: Take as many numbers of beans, blocks or coins as the right side of the equation shows and share them on the shapes to the left side of the equation.

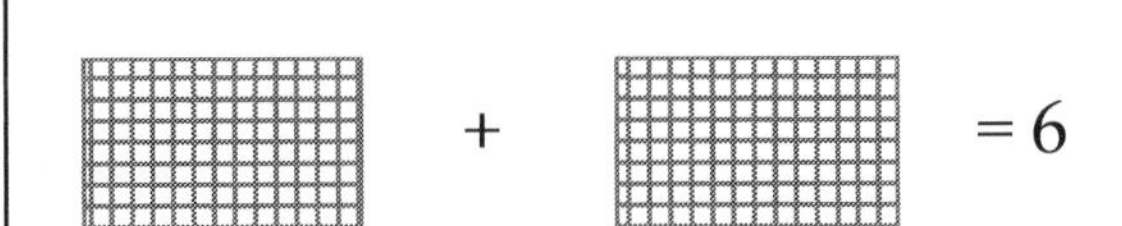

= ______ = ______

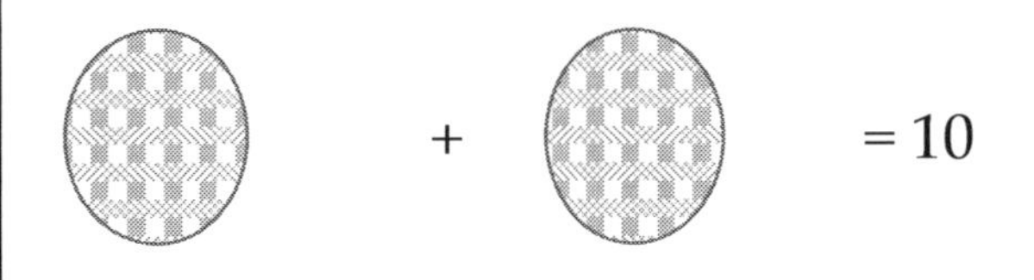

= ______ = ______

LESSON 9 – SKILL BUILDING

Visualize the numbers on the beam in your mind and draw it to represent the numbers given.

Monday

78 43 55 68 16

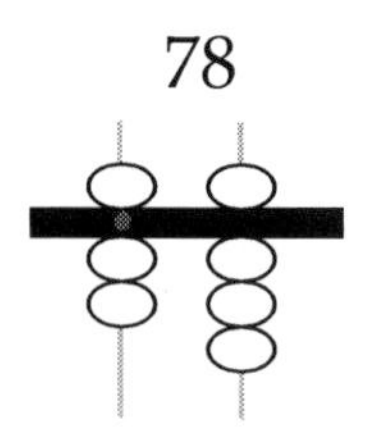
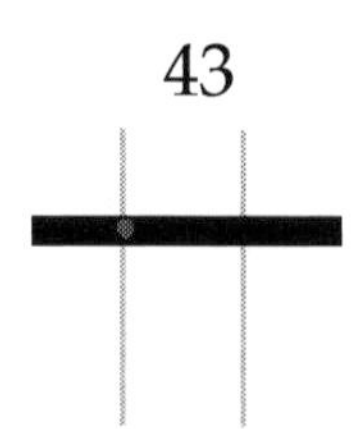
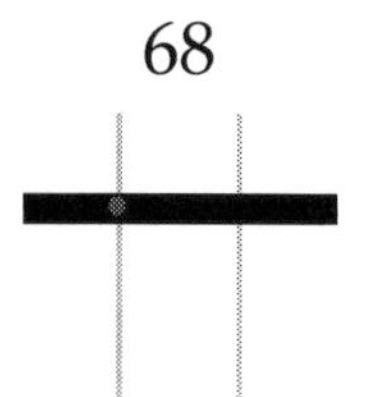
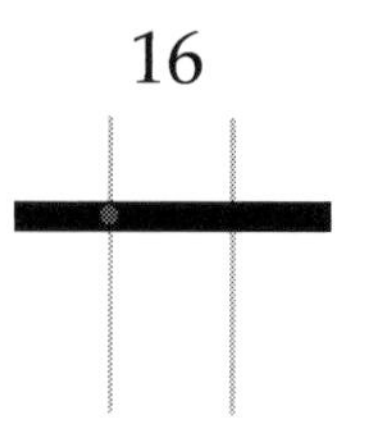

Tuesday

29 25 95 30 14

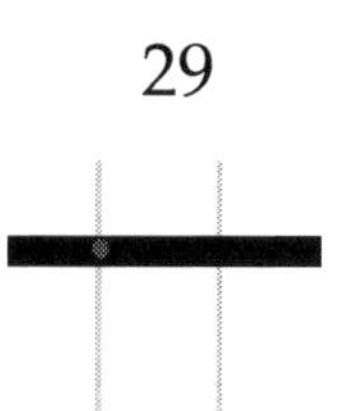
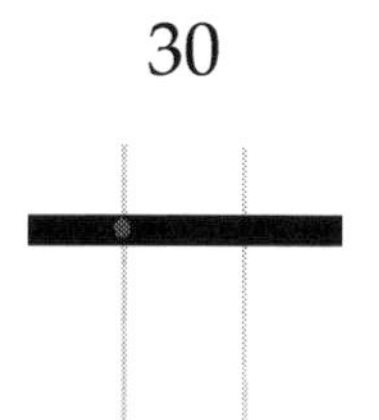
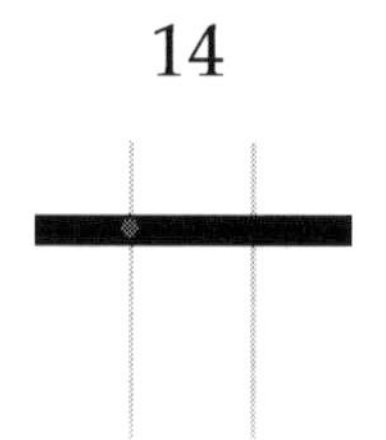

Wednesday

81 57 71 63 77

Thursday

21 92 86 36 19

Friday

39 78 46 90 05

WEEK 10 – LESSON 10 – INTRODUCING –1 CONCEPT

LESSON 10 – PRACTICE WORK

–1 = – 5 + 4

DAY 1 – MONDAY

Accuracy ________/32

Use Abacus

1	2	3	4	5	6	7	8	10:1
15	10	25	50	35	14	50	35	
30	25	60	- 10	- 21	31	05	- 31	
- 01	- 11	- 51	05	70	50	02	55	
10	20	- 10	- 01	- 54	- 21	- 11	- 18	

1	2	3	4	5	6	7	8	10:2
35	30	13	25	04	20	45	54	
- 11	10	- 10	23	50	25	50	- 12	
50	55	25	51	01	- 31	- 10	15	
10	- 21	- 08	- 88	- 10	05	- 11	- 22	

1	2	3	4	5	6	7	8	10:3
15	24	44	20	40	35	74	25	
- 11	21	51	24	55	11	01	20	
21	51	- 40	- 30	- 30	13	- 60	- 40	
51	01	- 11	01	- 11	- 09	- 01	- 01	

1	2	3	4	5	6	7	8	10:4
35	34	56	41	04	15	64	95	
- 21	11	03	13	51	70	11	- 41	
71	50	- 14	01	- 11	- 31	- 51	- 10	
04	- 31	- 41	- 15	- 44	- 10	15	51	

DAY 2 – TUESDAY

TIME: ________min ________sec Accuracy ______/16

1	2	3	4	5	6	7	8	
55	20	30	53	45	30	05	25	10:5
- 11	20	50	01	50	15	10	20	
- 21	05	15	- 10	- 31	- 11	- 11	- 31	
- 23	- 01	- 01	11	21	- 30	20	50	

1	2	3	4	5	6	7	8	
55	52	35	55	51	35	50	87	10:6
- 51	05	10	- 11	03	- 21	45	11	
40	- 15	- 21	01	- 10	31	- 51	- 43	
15	56	- 10	13	- 44	- 01	- 32	- 11	

DAY 3 – WEDNESDAY

TIME: ________min ________sec Accuracy ______/16

1	2	3	4	5	6	7	8	
15	65	24	74	35	43	51	54	10:7
- 11	- 11	21	01	- 01	11	25	01	
21	01	- 31	- 21	11	01	- 21	- 10	
01	- 15	01	- 13	14	- 11	- 11	- 40	

1	2	3	4	5	6	7	8	
34	10	75	64	04	44	15	46	10:8
11	04	- 21	21	51	11	50	12	
14	11	01	- 31	01	- 50	- 11	- 17	
- 02	- 01	- 15	- 13	- 50	- 01	- 11	18	

DAY 4 – THURSDAY

TIME: ________min ________sec Accuracy ______/16

1	2	3	4	5	6	7	8
85	59	54	58	64	25	75	35
- 31	- 11	31	- 17	11	- 01	- 21	51
01	50	- 51	16	- 51	20	05	- 66
- 15	- 46	15	- 15	70	10	- 14	- 20

10:9

1	2	3	4	5	6	7	8
42	65	54	49	55	85	50	87
02	30	11	- 04	10	- 01	05	- 22
- 31	- 01	- 10	10	- 11	- 10	- 11	- 11
66	- 73	- 11	- 01	- 11	- 63	- 33	- 11

10:10

DAY 5 – FRIDAY

TIME: ________min ________sec Accuracy ______/16

1	2	3	4	5	6	7	8
15	04	45	55	34	51	45	33
- 11	51	10	- 11	- 10	08	- 30	15
25	03	- 11	- 12	11	- 10	20	- 22
- 08	- 15	- 14	- 21	- 21	- 44	- 01	71

10:11

1	2	3	4	5	6	7	8
67	85	20	69	14	16	35	25
- 12	- 61	25	- 15	61	33	50	- 11
- 11	21	10	- 10	20	10	- 31	35
- 41	10	- 11	51	- 71	- 53	- 10	- 29

10:12

LESSON 10 – MIND MATH PRACTICE

Visualize

DAY 1 – MONDAY

Accuracy ______/10 ☆

1	2	3	4	5	6	7	8	9	10
05	04	06	05	53	54	04	35	44	25
- 01	01	- 01	- 01	- 02	01	51	- 01	01	- 01
- 03	04	- 01	- 01	- 10	- 10	- 10	- 02	- 10	10

10:13

DAY 2 – TUESDAY

Accuracy ______/10 ☆

1	2	3	4	5	6	7	8	9	10
					05	66	45	95	57
35	04	25	45	13	50	- 11	10	- 41	01
- 01	01	- 01	50	01	10	- 11	- 01	05	- 13
- 04	10	- 03	- 01	01	- 01	- 22	- 11	- 19	- 01

10:14

DAY 3 – WEDNESDAY

Accuracy ______/10 ☆

1	2	3	4	5	6	7	8	9	10
					43	15	22	45	25
53	12	02	64	25	11	- 01	22	10	- 01
01	10	42	01	- 11	01	50	11	- 10	20
- 11	05	11	- 51	- 11	- 10	01	- 01	- 41	11

10:15

DAY 4 – THURSDAY

Accuracy ______/10 ☆

1	2	3	4	5	6	7	8	9	10
52	10	36	13	01	55	45	45	55	75
- 11	55	11	- 03	51	- 11	- 31	- 21	- 10	- 61
50	10	- 20	25	01	- 33	10	50	- 01	50
- 80	- 51	- 07	- 01	- 10	- 11	10	- 70	- 22	- 04

10:16

DAY 5 – FRIDAY Accuracy ________/20

1	2	3	4	5	6	7	8	9	10
05	15	35	93	75	55	51	15	55	44
- 01	- 11	- 01	- 73	- 21	- 10	03	- 11	- 11	11
01	05	60	05	- 10	50	- 11	20	11	- 01
04	- 06	- 20	- 01	- 44	- 01	- 40	20	04	- 13

10:17

1	2	3	4	5	6	7	8	9	10
45	14	45	44	52	34	40	52	25	65
- 01	11	- 01	11	- 10	50	03	12	- 21	30
10	- 01	- 02	03	02	- 30	01	- 13	40	- 90
- 04	- 22	10	- 10	01	- 10	- 11	- 10	10	- 01

10:18

SHAPE PUZZLE

Find out what each shape is worth.

Suggestion: Take as many numbers of beans, blocks or coins as the right side of the equation shows and share them on the shapes to the left side of the equation.

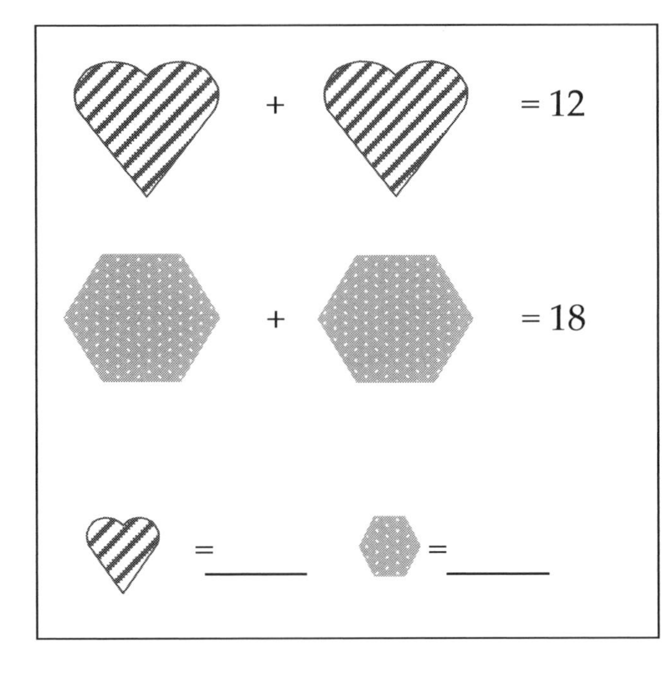

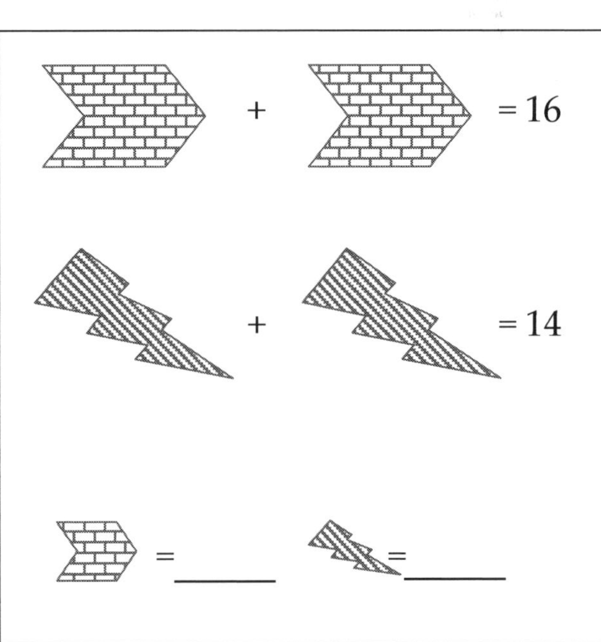

LESSON 10 – SKILL BUILDING

Visualize the numbers on the beam in your mind and draw it to represent the numbers given.

78	58	39	24	66	Monday
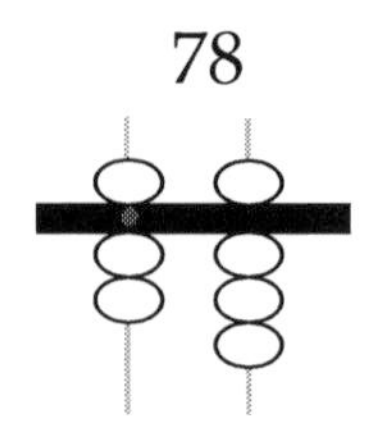	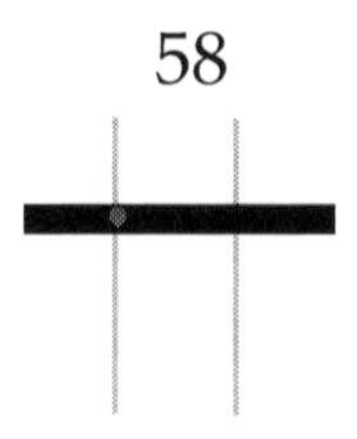	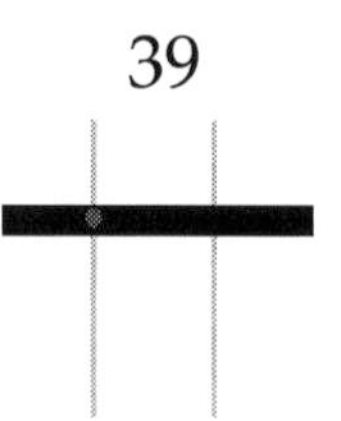	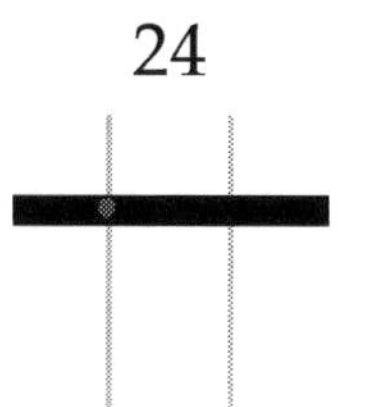	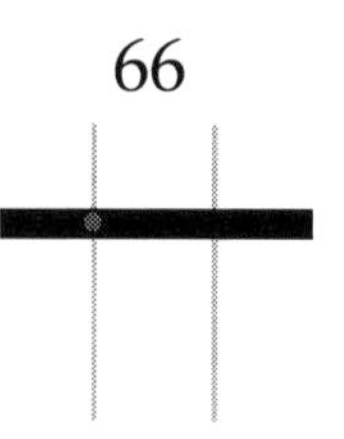	
73	84	96	05	18	Tuesday
50	33	61	26	48	Wednesday
62	51	04	59	74	Thursday
42	17	08	32	12	Friday

ANSWER KEY

PLACE VALUE OF NUMBERS PRACTICE

42 =	4 tens	2 ones
90 =	9 tens	0 ones
19 =	1 tens	9 ones
24 =	2 tens	4 ones
11 =	1 tens	1 ones
12 =	1 tens	2 ones
32 =	3 tens	2 ones
64 =	6 tens	4 ones
73 =	7 tens	3 ones
59 =	5 tens	9 ones
15 =	1 tens	5 ones
45 =	4 tens	5 ones
83 =	8 tens	3 ones
61 =	6 tens	1 ones
05 =	0 tens	5 ones
37 =	3 tens	7 ones
06 =	0 tens	6 ones
29 =	2 tens	9 ones
92 =	9 tens	2 ones
56 =	5 tens	6 ones
91 =	9 tens	1 ones
07 =	0 tens	7 ones
79 =	7 tens	9 ones
54 =	5 tens	4 ones
68 =	6 tens	8 ones
20 =	2 tens	0 ones

What place value is 2 in the following numbers? Write the name of the place value on the given line.

21 → Tens
23 → Tens
302→ Ones
12 → Ones
62 → Ones
126→ Tens
02 → Ones
24 → Tens
182→ Ones
32 → Ones
52 → Ones
428→ Tens

What place value is 5 in the following numbers? Write the name of the place value on the given line.

51 → Tens
56 → Tens
605 → Ones
35 → Ones
85 → Ones
750 → Tens
05 → Ones
57 → Tens
185 → Ones
50 → Tens
95 → Ones
451 → Tens

What place value is 7 in the following numbers? Write the name of the place value on the given line.

37 → Ones
47 → Ones
507→ Ones
75 → Tens
07 → Ones
375→ Tens
17 → Tens
79 → Tens
474→ Tens
67 → Ones
27 → Ones
187→ Ones

WEEK 1 – PRACTICE WORK

DAY 1 – MONDAY

1	2	3	4	5	6	7	8	9	10	
03	04	03	03	04	02	04	02	14	22	1:1

1	2	3	4	5	6	7	8	9	10	
23	12	44	03	01	04	40	20	30	40	1:2

1	2	3	4	5	6	7	8	9	10	
10	30	40	10	03	02	04	03	01	00	1:3

1	2	3	4	5	6	7	8	9	10	
40	10	00	30	10	00	20	00	00	00	1:4

DAY 2 – TUESDAY

1	2	3	4	5	6	7	8	
02	30	11	10	20	00	21	32	1:5

1	2	3	4	5	6	7	8	
24	11	01	32	14	11	31	00	1:6

DAY 3 – WEDNESDAY

1	2	3	4	5	6	7	8	
14	00	43	00	13	33	43	43	1:7

1	2	3	4	5	6	7	8	
34	01	10	00	12	34	01	12	1:8

DAY 4 – THURSDAY

1	2	3	4	5	6	7	8	
40	03	43	11	10	44	12	03	1:9

1	2	3	4	5	6	7	8	
44	44	34	44	31	12	32	41	1:10

DAY 5 – FRIDAY

1	2	3	4	5	6	7	8	
24	34	44	44	32	22	22	32	1:11

1	2	3	4	5	6	7	8	
12	23	32	01	12	43	03	24	1:12

WEEK 2 – PRACTICE WORK

DAY 1 – MONDAY

1	2	3	4	5	6	7	8	9	10	
40	55	50	34	20	35	13	45	44	13	2:1

1	2	3	4	5	6	7	8	9	10	
10	11	45	35	41	05	13	00	04	55	2:2

1	2	3	4	5	6	7	8	9	10	
50	25	52	05	01	24	52	51	55	53	2:3

1	2	3	4	5	6	7	8	9	10	
44	05	12	50	30	15	42	21	25	52	2:4

DAY 2 – TUESDAY

1	2	3	4	5	6	7	8	
31	15	35	15	51	31	03	50	2:5

1	2	3	4	5	6	7	8	
12	05	22	53	20	35	05	00	2:6

DAY 3 – WEDNESDAY

1	2	3	4	5	6	7	8	
45	13	25	15	55	53	12	55	2:7

1	2	3	4	5	6	7	8	
54	05	25	35	02	45	01	35	2:8

DAY 4 – THURSDAY

1	2	3	4	5	6	7	8	
30	35	54	32	45	55	54	00	2:9

1	2	3	4	5	6	7	8	
15	55	02	22	33	54	55	51	2:10

DAY 5 – FRIDAY

1	2	3	4	5	6	7	8	
30	05	40	50	45	11	50	05	2:11

1	2	3	4	5	6	7	8	
23	45	45	15	02	55	53	10	2:12

WEEK 3 – PRACTICE WORK

DAY 1 – MONDAY

1	2	3	4	5	6	7	8	9	10	
06	06	07	07	01	05	07	05	07	07	3:1

1	2	3	4	5	6	7	8	9	10	
70	60	60	60	70	07	07	37	47	01	3:2

1	2	3	4	5	6	7	8	9	10	
51	74	12	70	60	70	52	16	37	06	3:3

1	2	3	4	5	6	7	8	9	10	
70	46	50	10	07	01	26	50	50	00	3:4

DAY 2 – TUESDAY

1	2	3	4	5	6	7	8	
36	24	06	60	47	50	50	56	3:5

1	2	3	4	5	6	7	8	
56	06	26	60	76	77	76	67	3:6

DAY 3 – WEDNESDAY

1	2	3	4	5	6	7	8	
36	77	41	37	07	63	06	76	3:7

1	2	3	4	5	6	7	8	
50	07	21	77	56	62	52	50	3:8

DAY 4 – THURSDAY

1	2	3	4	5	6	7	8	
44	36	73	52	37	26	64	76	3:9

1	2	3	4	5	6	7	8	
77	13	71	26	77	34	76	05	3:10

DAY 5 – FRIDAY

1	2	3	4	5	6	7	8	
21	16	67	67	31	73	67	50	3:11

1	2	3	4	5	6	7	8	
76	30	67	61	73	57	64	77	3:12

WEEK 4 – PRACTICE WORK

DAY 1 – MONDAY

1	2	3	4	5	6	7	8	9	10	
08	80	09	09	90	50	80	49	48	47	4:1

1	2	3	4	5	6	7	8	9	10	
08	08	09	09	13	70	90	90	50	80	4:2

1	2	3	4	5	6	7	8	9	10	
08	09	06	65	87	68	90	50	40	00	4:3

1	2	3	4	5	6	7	8	9	10	
24	89	97	73	98	21	15	35	77	71	4:4

DAY 2 – TUESDAY

1	2	3	4	5	6	7	8	
08	05	06	90	10	08	23	32	4:5

1	2	3	4	5	6	7	8	
16	87	16	49	79	79	13	87	4:6

DAY 3 – WEDNESDAY

1	2	3	4	5	6	7	8	
08	00	80	09	90	24	15	07	4:7

1	2	3	4	5	6	7	8	
70	89	79	00	06	71	11	98	4:8

DAY 4 – THURSDAY

1	2	3	4	5	6	7	8	
09	80	20	10	78	69	60	78	4:9

1	2	3	4	5	6	7	8	
95	85	08	76	94	72	71	94	4:10

DAY 5 – FRIDAY

1	2	3	4	5	6	7	8	
58	90	39	69	89	90	78	52	4:11

1	2	3	4	5	6	7	8	
79	17	58	98	87	67	36	12	4:12

WEEK 5 – PRACTICE WORK

DAY 1 – MONDAY

1	2	3	4	5	6	7	8	5:1
49	80	07	90	60	97	48	89	

1	2	3	4	5	6	7	8	5:2
99	54	11	02	02	98	86	48	

DAY 2 – TUESDAY

1	2	3	4	5	6	7	8	5:3
94	34	05	92	26	97	60	47	

1	2	3	4	5	6	7	8	5:4
88	71	27	98	53	89	38	74	

DAY 3 – WEDNESDAY

1	2	3	4	5	6	7	8	5:5
31	13	00	24	99	76	94	52	

1	2	3	4	5	6	7	8	5:6
27	29	30	48	81	62	50	00	

DAY 4 – THURSDAY

1	2	3	4	5	6	7	8	5:7
03	12	89	02	47	44	15	53	

1	2	3	4	5	6	7	8	5:8
99	71	45	86	52	97	72	01	

DAY 5 – FRIDAY

1	2	3	4	5	6	7	8	5:9
83	34	87	79	11	44	61	71	

1	2	3	4	5	6	7	8	5:10
71	25	22	27	99	13	02	98	

WEEK 5 – MIND MATH PRACTICE WORK

DAY 1 – MONDAY

1	2	3	4	5	6	7	8	9	10	5:11
03	03	04	03	01	02	04	02	04	01	

1	2	3	4	5	6	7	8	9	10	5:12
20	40	20	30	10	04	40	20	22	20	

1	2	3	4	5	6	7	8	9	10	5:13
42	31	10	23	32	03	32	13	22	21	

1	2	3	4	5	6	7	8	9	10	5:14
25	25	45	15	05	00	50	52	25	51	

DAY 2 – TUESDAY

1	2	3	4	5	6	7	8	9	10	5:15
33	44	33	33	10	32	35	25	15	45	

DAY 3 – WEDNESDAY

1	2	3	4	5	6	7	8	9	10	5:16
44	02	12	30	05	05	50	05	10	00	

DAY 4 – THURSDAY

1	2	3	4	5	6	7	8	9	10	5:17
00	22	15	10	10	02	05	45	10	53	

DAY 5 – FRIDAY

1	2	3	4	5	6	7	8	9	10	5:18
51	51	51	55	45	35	00	25	30	00	

1	2	3	4	5	6	7	8	9	10	5:19
54	35	05	15	45	10	33	22	35	25	

WEEK 6 – PRACTICE WORK

DAY 1 – MONDAY

1	2	3	4	5	6	7	8	
90	54	03	77	16	43	46	80	6:1

1	2	3	4	5	6	7	8	
79	00	20	00	63	33	48	05	6:2

DAY 2 – TUESDAY

1	2	3	4	5	6	7	8	
98	93	81	88	52	84	00	02	6:3

1	2	3	4	5	6	7	8	
26	72	19	59	57	00	49	20	6:4

DAY 3 – WEDNESDAY

1	2	3	4	5	6	7	8	
84	22	48	76	42	79	94	98	6:5

1	2	3	4	5	6	7	8	
88	11	85	11	97	16	87	74	6:6

DAY 4 – THURSDAY

1	2	3	4	5	6	7	8	
97	36	01	05	10	34	30	00	6:7

1	2	3	4	5	6	7	8	
20	15	97	01	04	99	80	97	6:8

DAY 5 – FRIDAY

1	2	3	4	5	6	7	8	
03	78	13	62	37	02	51	12	6:9

1	2	3	4	5	6	7	8	
79	10	00	01	12	99	39	94	6:10

WEEK 6 – MIND MATH PRACTICE WORK

DAY 1 – MONDAY

1	2	3	4	5	6	7	8	9	10	
01	07	05	07	05	06	03	00	12	16	6:11

1	2	3	4	5	6	7	8	9	10	
10	60	50	70	60	10	10	60	10	70	6:12

1	2	3	4	5	6	7	8	9	10	
61	75	04	56	24	23	77	60	67	14	6:13

1	2	3	4	5	6	7	8	9	10	
06	12	56	75	03	60	01	52	66	32	6:14

DAY 2 – TUESDAY

1	2	3	4	5	6	7	8	9	10	
01	00	06	00	27	20	20	01	55	46	6:15

DAY 3 – WEDNESDAY

1	2	3	4	5	6	7	8	9	10	
40	20	60	00	75	52	65	26	20	40	6:16

DAY 4 – THURSDAY

1	2	3	4	5	6	7	8	9	10	
35	25	13	70	57	00	46	04	57	11	6:17

DAY 5 – FRIDAY

1	2	3	4	5	6	7	8	9	10	
72	20	72	25	74	61	60	70	51	72	6:18

1	2	3	4	5	6	7	8	9	10	
16	27	06	51	63	23	77	00	77	64	6:19

WEEK 7 – PRACTICE WORK

DAY 1 – MONDAY

1	2	3	4	5	6	7	8	
94	94	88	79	64	30	24	41	7:1

1	2	3	4	5	6	7	8	
69	27	86	29	32	85	48	86	7:2

DAY 2 – TUESDAY

1	2	3	4	5	6	7	8	
89	27	72	10	22	01	16	09	7:3

1	2	3	4	5	6	7	8	
19	28	21	82	00	08	28	69	7:4

DAY 3 – WEDNESDAY

1	2	3	4	5	6	7	8	
97	19	12	98	51	99	11	02	7:5

1	2	3	4	5	6	7	8	
51	28	58	34	00	89	89	15	7:6

DAY 4 – THURSDAY

1	2	3	4	5	6	7	8	
86	12	98	15	82	78	84	70	7:7

1	2	3	4	5	6	7	8	
80	56	57	97	96	47	12	16	7:8

DAY 5 – FRIDAY

1	2	3	4	5	6	7	8	
32	69	00	88	87	92	99	97	7:9

1	2	3	4	5	6	7	8	
96	00	81	56	96	48	00	60	7:10

WEEK 7 – MIND MATH PRACTICE WORK

DAY 1 – MONDAY

1	2	3	4	5	6	7	8	9	10	
07	08	09	08	05	07	08	05	04	02	7:11

1	2	3	4	5	6	7	8	9	10	
00	40	90	60	90	00	90	45	91	95	7:12

1	2	3	4	5	6	7	8	9	10	
80	30	60	80	80	63	76	59	92	80	7:13

1	2	3	4	5	6	7	8	9	10	
99	88	79	77	59	85	39	18	55	93	7:14

DAY 2 – TUESDAY

1	2	3	4	5	6	7	8	9	10	
28	49	56	96	68	85	95	90	79	62	7:15

DAY 3 – WEDNESDAY

1	2	3	4	5	6	7	8	9	10	
09	08	07	95	96	68	59	80	80	76	7:16

DAY 4 – THURSDAY

1	2	3	4	5	6	7	8	9	10	
08	19	67	10	40	80	91	97	80	72	7:17

DAY 5 – FRIDAY

1	2	3	4	5	6	7	8	9	10	
28	19	58	92	78	70	91	85	02	57	7:18

1	2	3	4	5	6	7	8	9	10	
84	49	58	91	18	50	13	09	19	89	7:19

WEEK 8 – PRACTICE WORK

DAY 1 – MONDAY

1	2	3	4	5	6	7	8	
25	48	28	99	54	86	63	78	8:1

1	2	3	4	5	6	7	8	
97	71	45	48	69	94	79	02	8:2

DAY 2 – TUESDAY

1	2	3	4	5	6	7	8	
46	58	38	67	47	15	12	14	8:3

1	2	3	4	5	6	7	8	
05	36	73	77	06	73	59	77	8:4

DAY 3 – WEDNESDAY

1	2	3	4	5	6	7	8	
37	00	71	60	71	67	98	00	8:5

1	2	3	4	5	6	7	8	
89	94	58	85	00	54	45	19	8:6

DAY 4 – THURSDAY

1	2	3	4	5	6	7	8	
89	27	85	10	12	99	00	95	8:7

1	2	3	4	5	6	7	8	
79	55	86	97	60	47	03	02	8:8

DAY 5 – FRIDAY

1	2	3	4	5	6	7	8	
30	21	21	37	41	62	49	17	8:9

1	2	3	4	5	6	7	8	
49	87	27	10	07	49	00	13	8:10

WEEK 8 – MIND MATH PRACTICE WORK

DAY 1 – MONDAY

1	2	3	4	5	6	7	8	9	10	
71	81	05	77	52	91	18	41	57	50	8:11

DAY 2 – TUESDAY

1	2	3	4	5	6	7	8	9	10	
74	40	61	83	04	44	54	01	00	15	8:12

DAY 3 – WEDNESDAY

1	2	3	4	5	6	7	8	9	10	
71	45	63	37	59	81	49	28	39	61	8:13

DAY 4 – THURSDAY

1	2	3	4	5	6	7	8	9	10	
79	00	01	65	35	96	01	00	99	09	8:14

DAY 5 – FRIDAY

1	2	3	4	5	6	7	8	9	10	
20	08	38	61	58	12	07	25	49	91	8:15

1	2	3	4	5	6	7	8	9	10	
42	79	55	00	56	08	24	90	18	17	8:16

WEEK 9 – PRACTICE WORK

DAY 1 – MONDAY

1	2	3	4	5	6	7	8	
45	54	55	50	55	46	55	45	9:1

1	2	3	4	5	6	7	8	
10	06	25	78	85	57	55	23	9:2

1	2	3	4	5	6	7	8	
50	00	55	58	57	50	65	55	9:3

1	2	3	4	5	6	7	8	
01	60	95	75	24	99	99	59	9:4

DAY 2 – TUESDAY

1	2	3	4	5	6	7	8	
45	45	04	95	75	75	60	25	9:5

1	2	3	4	5	6	7	8	
99	75	99	95	55	53	65	00	9:6

DAY 3 – WEDNESDAY

1	2	3	4	5	6	7	8	
96	93	25	16	86	20	09	12	9:7

1	2	3	4	5	6	7	8	
57	49	96	96	05	15	60	65	9:8

DAY 4 – THURSDAY

1	2	3	4	5	6	7	8	
65	40	50	69	79	50	51	05	9:9

1	2	3	4	5	6	7	8	
02	06	05	05	20	31	10	96	9:10

DAY 5 – FRIDAY

1	2	3	4	5	6	7	8	
57	58	55	95	58	58	01	59	9:11

1	2	3	4	5	6	7	8	
23	84	86	98	95	50	56	83	9:12

WEEK 9 – MIND MATH PRACTICE WORK

DAY 1 – MONDAY

1	2	3	4	5	6	7	8	9	10	
20	29	22	45	15	30	15	25	58	15	9:13

DAY 2 – TUESDAY

1	2	3	4	5	6	7	8	9	10	
35	27	25	16	52	50	11	56	10	20	9:14

DAY 3 – WEDNESDAY

1	2	3	4	5	6	7	8	9	10	
45	00	23	35	26	56	02	28	70	55	9:15

DAY 4 – THURSDAY

1	2	3	4	5	6	7	8	9	10	
61	02	45	55	72	00	80	06	00	05	9:16

DAY 5 – FRIDAY

1	2	3	4	5	6	7	8	9	10	
52	05	15	58	59	50	35	99	50	00	9:17

1	2	3	4	5	6	7	8	9	10	
00	05	25	78	65	95	43	55	65	51	9:18

WEEK 9 – SHAPE PUZZLES

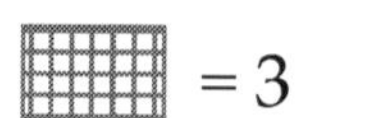 = 3

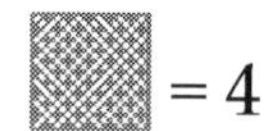 = 4

= 2

= 5

WEEK 10 – PRACTICE WORK

DAY 1 – MONDAY

1	2	3	4	5	6	7	8	
54	44	24	44	30	74	46	41	10:1

1	2	3	4	5	6	7	8	
84	74	20	11	45	19	74	35	10:2

1	2	3	4	5	6	7	8	
76	97	44	15	54	50	14	04	10:3

1	2	3	4	5	6	7	8	
89	64	04	40	00	44	39	95	10:4

DAY 2 – TUESDAY

1	2	3	4	5	6	7	8	
00	44	94	55	85	04	24	64	10:5

1	2	3	4	5	6	7	8	
59	98	14	58	00	44	12	44	10:6

DAY 3 – WEDNESDAY

1	2	3	4	5	6	7	8	
26	40	15	41	59	44	44	05	10:7

1	2	3	4	5	6	7	8	
57	24	40	41	06	04	43	59	10:8

DAY 4 – THURSDAY

1	2	3	4	5	6	7	8	
40	52	49	42	94	54	45	00	10:9

1	2	3	4	5	6	7	8	
79	21	44	54	43	11	11	43	10:10

DAY 5 – FRIDAY

1	2	3	4	5	6	7	8	
21	43	30	11	14	05	34	97	10:11

1	2	3	4	5	6	7	8	
03	55	44	95	24	06	44	20	10:12

WEEK 10 – MIND MATH PRACTICE WORK

DAY 1 – MONDAY

1	2	3	4	5	6	7	8	9	10	
01	09	04	03	41	45	45	32	35	34	10:13

DAY 2 – TUESDAY

1	2	3	4	5	6	7	8	9	10	
30	15	21	94	15	64	22	43	40	44	10:14

DAY 3 – WEDNESDAY

1	2	3	4	5	6	7	8	9	10	
43	27	55	14	03	45	65	54	04	55	10:15

DAY 4 – THURSDAY

1	2	3	4	5	6	7	8	9	10	
11	24	20	34	43	00	34	04	22	60	10:16

DAY 5 – FRIDAY

1	2	3	4	5	6	7	8	9	10	
09	03	74	24	00	94	03	44	59	41	10:17

1	2	3	4	5	6	7	8	9	10	
50	02	52	48	45	44	33	41	54	04	10:18

WEEK 10 – SHAPE PUZZLES

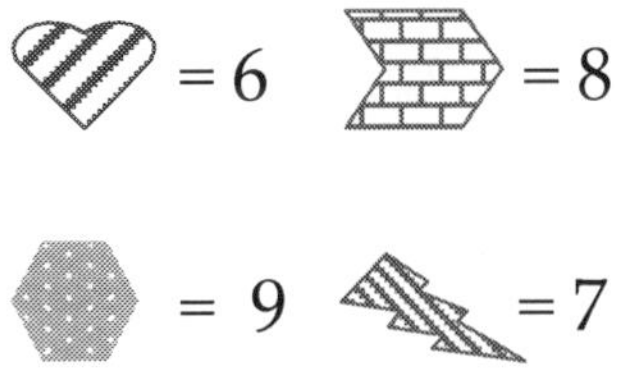

= 6 = 8

= 9 = 7

ABOUT SAI SPEED MATH ACADEMY

One subject that is very important for success in this world, along with being able to read and write, is the knowledge of numbers. Math is one subject which requires proficiency from anyone who wants to achieve something in life. A strong foundation and a basic understanding of math is a must to mastering higher levels of math.

We, the family, best friends, and parents of children in elementary school, early on discovered that what our children were learning at school was not enough for them to master the basics of math. Teachers at school, with the resources they had, did the best they could. But, as parents, we had to do more to help them understand the relationship between numbers and basic functions of adding, subtracting, multiplying and dividing. Also, what made us cringe is the fact that our children's attitude towards more complex math was to say, "Oh, we are allowed to use a calculator in class". This did not sit well with us. Even though we did not have a specific system that we followed, each of us could do basic calculations in our minds without looking for a calculator. So, this made us want to do more for our children.

We started to look into the various methods that were available in the marketplace to help our children understand basic math and reduce their dependency on calculators. We came across soroban, a wonderful calculating tool from Japan. Soroban perfectly fits with the base-10 number system used at present and provides a systematic method to follow while calculating in one's mind.

This convinced us and within a short time we were able to work with fluency on the tool. The next step was to introduce it to our children, which we thought was going to be an easy task. It, however, was not. It was next to impossible to find the resources or the curriculum to help us introduce the tool in the correct order. Teaching all the concepts in one sitting and expecting children to apply them to the set of problems we gave them only made them push away the tool in frustration.

However, help comes to those who ask, and to those who are willing to work to achieve their goals. We came across a soroban teacher who helped us by giving us ideas and an outline of how soroban should be introduced. But, we still needed an actual worksheet to give our children to practice on. That is when we decided to come up with practice worksheets of our own design for our kids.

Slowly and steadily, practicing with the worksheets that we developed, our children started to get the idea and loved what they could do with a soroban. Soon we realized that they were better with mind math than we were.

Today, 6 years later, all our kids have completed their soroban training and are reaping the benefits of the hard work that they did over the years.

Now, although very happy, we were humbled at the number of requests we got from parents who wanted to know more about our curriculum. We had no way to share our new knowledge with them.

Now, through the introduction of our instruction book and workbooks, that has changed. We want to share everything we know with all the dedicated parents who are interested in teaching soroban to their children. This is our humble attempt to bring a systematic instruction manual and corresponding workbook to help introduce your children to soroban.

What started as a project to help our kids has grown over the years and we are fortunate to say that a number of children have benefitted learning with the same curriculum that we developed for our children.

Thank you for choosing our system to enhance your children's mathematical skills.

We love working on soroban and hope you do too!

List of SAI Speed Math Academy Publications

LEVEL – 1

Abacus Mind Math Instruction Book Level – 1: Step by Step Guide to Excel at Mind Math with Soroban, a Japanese Abacus

ISBN-13: 978-1941589007

Abacus Mind Math Level – 1 Workbook 1 of 2: Excel at Mind Math with Soroban, a Japanese Abacus

ISBN-13: 978-1941589014

Abacus Mind Math Level – 1 Workbook 2 of 2: Excel at Mind Math with Soroban, a Japanese Abacus

ISBN-13: 978-1941589021

LEVEL – 2

Abacus Mind Math Instruction Book Level – 2: Step by Step Guide to Excel at Mind Math with Soroban, a Japanese Abacus

ISBN-13: 978-1941589038

Abacus Mind Math Level – 2 Workbook 1 of 2: Excel at Mind Math with Soroban, a Japanese Abacus

ISBN-13: 978-1941589045

Abacus Mind Math Level – 2 Workbook 2 of 2: Excel at Mind Math with Soroban, a Japanese Abacus

ISBN-13: 978-1941589052

LEVEL – 3

Abacus Mind Math Instruction Book Level – 3: Step by Step Guide to Excel at Mind Math with Soroban, a Japanese Abacus

ISBN-13: 9781941589069

Abacus Mind Math Level – 3 Workbook 1 of 2: Excel at Mind Math with Soroban, a Japanese Abacus

ISBN-13: 9781941589076

Abacus Mind Math Level – 3 Workbook 2 of 2: Excel at Mind Math with Soroban, a Japanese Abacus

ISBN-13: 9781941589083

Made in the USA
San Bernardino, CA
13 July 2015